Eberstadt and Forrestal

TEXAS A&M UNIVERSITY MILITARY HISTORY SERIES

21

Eberstadt and Forrestal

A NATIONAL SECURITY PARTNERSHIP, 1909–1949

Jeffery M. Dorwart

TEXAS A&M UNIVERSITY PRESS
COLLEGE STATION

First Edition

The paper used in this book meets the minimum requirements of the American National Standard for Permanence of Paper for Printed Library Materials, Z39.48-1984. Binding materials have been chosen for durability. ∞

LIBRARY OF CONGRESS CATALOGING-IN-PUBLICATION DATA

Dorwart, Jeffery M., 1944–
Eberstadt and Forrestal : a national security partnership, 1909–1949 / Jeffery M. Dorwart. – 1st ed.
p. cm. – (Texas A&M University military history series ; 21)
Includes bibliographical references and index.
ISBN 0-89096-469-6
1. Industrial mobilization–United States. 2. United States–History, Military–20th century. 3. World War, 1939–1945–Economic aspects. 4. Eberstadt, Ferdinand. 5. Forrestal, James, 1892–1949. I. Title. II. Series.
UA18.U5D67 1991
355.2′6′0973–dc20 91-8769
CIP

Contents

Preface and Acknowledgments

The National Security Act of 1947 has been blamed for everything from the creation of a military-industrial complex to the onset of the Cold War, from CIA excesses to "chilling" National Security Council conduct, and from bloated Pentagon budgets to bitter feuds among the armed services. In 1983 one critic wrote that "the tragedy of the National Security Act of 1947 was that it allowed perpetuation of four separate air forces," which meant that the public was "paying an enormous price in duplication for this flaw." Another critic of the American national security establishment, journalist Arthur T. Hadley claimed that the 1947 law contributed to American military failures in Vietnam, Laos, Cambodia, Cuba, Nicaragua, Angola, and Libya. Hadley insisted that the chaotic creation of 1947 caused the disastrous Iranian rescue mission of April, 1980. Hadley condemned the flawed national security organization for making the United States into a "straw giant," with its military weakened by interservice rivalries, devoid of a strategic plan, strangled by a ponderous bureaucracy, and divorced from civilian leaders and public.[1]

Some of the criticism has come from those who advocate a particular form of American defense organization or oppose American involvement in conflicts such as the Vietnam War. Their views cannot be dismissed simply as the parochial perspective of special service interests or the protestations of antiwar critics. Over the years, American national security organization has not always operated efficiently or economically and has suffered a number of serious setbacks and failures. Those problems prompt numerous questions: How did the United States build such a flawed system of national security? Who was responsible for its creation? Why was the National Security Act a chaotic piece of legislation and what form of organization did it construct? What did the law actually do?

The study presented here shows that the origins and creation of a large-scale organization for American national security were complex and convoluted, arising from a number of interrelated events and sources. The development of national security structures was rooted in the larger process of building a new American administrative state during the first part of the twentieth century and was influenced by the mobilization for both world wars. National security organization received impetus from Soviet-American confrontation in the Cold War and from the developments in military technology, particularly the development of air power and the atomic bomb.

A number of creative individuals were behind the creation of the national security organization. Elihu Root and other turn-of-the-century organizational elite laid the foundations for national security. Bernard Baruch and Woodrow Wilson's war mobilizers contributed their experiences and structures from World War I. Herbert Hoover's theories of an associative state and the management principles of Owen Young and other corporate liberals found expression in national security organization. Army officers John M. Palmer and George C. Marshall stimulated unification of the military and provided the catalyst for national security organization. Harold Smith, budget director for FDR and Truman adopted the politics of national defense reorganization to strengthen the presidency. There were many others involved, but the concepts and forms laid out in the National Security Act of 1947 derived most directly from the activities of two men, Ferdinand Eberstadt and James V. Forrestal. This book is about these two institutional innovators and the organization that they built.

The preparation of this study received help from a number of sources. A grant-in-aid from the Harry S. Truman Library Institute helped finance early research. The staffs at the Truman and Franklin D. Roosevelt presidential libraries were most helpful. The staff at the George C. Marshall Research Library photocopied documents for me that related to Marshall's unification plan. Dean Allard and his staff at the Operational Archives, Naval History Division, contributed their usual courteous and efficient assistance. Ben Frank showed me the Marine Corps oral history collection on unification. Nancy Bressler and her staff at the Seeley Mudd Manuscript Library at Princeton University provided prompt, steady assistance and secured permission for me to cite materials from the James Forrestal Papers. The entire staff of the Rutgers University Library in Camden contributed to the acquisition of books on interlibrary loan.

A number of individuals helped in research, including my former students Stephen Walter and Claire Myer Schwarz. My Rutgers colleagues Andrew Lees and Rodney P. Carlisle read various drafts of the manuscript. Carlisle discussed every revision with me. Lawrence Legere and William B. Huie answered my questions about unification. Milton Katz shared his memories about the Eberstadt study group, and Frederick Eberstadt showed great patience and candor in discussing with me his father's personality and relationships.

This work would have been nearly impossible without the assistance of Calvin P. Christman, an Eberstadt scholar who shared both his enthusiasm for this project and his extensive collection of correspondence about Eberstadt.

Special thanks go to Loretta Carlisle, who typed and meticulously copyedited every version of the manuscript and constantly encouraged me to see the project through to its completion. Special thanks, too, go to Nel, Catherine, Karl, and Candace for tolerating my bad temper during the stressful hours spent revising this text.

Eberstadt and Forrestal

1. Eberstadt, Forrestal, and the Good Man Concept

Ferdinand Eberstadt and James Forrestal slipped into Washington nearly unnoticed among the hundreds of businessmen who assembled in 1940 to prepare the United States for war. These two wealthy investment bankers were intimate friends, and they moved quickly to the center of the war mobilization program and into the process of building institutional structures for national security. Over the next nine years, they became major architects of the modern American national security establishment.

Forrestal served between 1940 and 1949 as undersecretary and secretary of the navy and as the first secretary of defense. Eberstadt's official positions were more modest than those held by Forrestal but more enduring and instrumental in determining the shape of American national security institutions. During the second world war he became chairman of the Army Navy Munitions Board (ANMB) and vice-chairman of the War Production Board (WPB). After the war, Eberstadt headed the Hoover Commission Task Force on National Security. More significantly, Eberstadt designed the Controlled Materials Plan, which rationalized the nation's war resources, and prepared a blueprint for military unification that became the National Security Act of 1947. The National Security Council, National Security Resources Board, Central Intelligence Agency, and Department of Defense were his "offspring," he told Secretary of Defense Robert McNamara in 1967.[1]

Eberstadt did not greatly exaggerate his contribution. His ideas and actions made him perhaps the single most important organizer of the American national security establishment. However, Eberstadt's accomplishments depended upon his close relationship with Forrestal. Without Forrestal, Eberstadt's impact would have been no greater than that of dozens of other management and organizational experts who held key mobilization and defense posts between 1940 and 1949. Forrestal's positions in the Roosevelt and Truman administrations gave Eberstadt immediate entrée to power. Forrestal introduced his best friend to the top civilian and military commanders of the American war machine. He connected Eberstadt to Washington's most important political and congressional leaders. Through Forrestal, Eberstadt became an intermediary between public and private sectors: government and finance, military and industry. Unrestrained by loyalty to either Roosevelt or Truman, Eberstadt battled in the open and behind the scenes to graft theories, methods, and structures from the business world onto public institutions.

Eberstadt and Forrestal needed each other to become effective organizers of national security institutions. They formed an inseparable team and a type of public-private partnership. Forrestal resided on the inside as a top public official with decision-making authority, while Eberstadt operated mostly on the outside as an expert free to move across institutional boundaries. They were two creative individuals who at critical moments held positions of power and leadership essential for guiding institutional change and building interorganizational structures for national security. Forrestal brought Eberstadt into the process because he depended upon his friend for advice and support. For twenty-five years, Eberstadt had helped Forrestal solve legal, business, and personal problems, and Forrestal's government job only increased his reliance on his intimate friend. As a defense official between 1940 and 1949, Forrestal confronted a seemingly endless succession of crises brought about by production bottlenecks, bitter bureaucratic power struggles, and interorganizational conflicts. Each new crisis led to that inevitable late-night telephone call from a worried Forrestal asking Eberstadt to hurry down to Washington to bail him out of yet another organizational tangle.[2]

Forrestal heeded Eberstadt's insight and ability to solve problems. Eberstadt moved directly to the heart of a matter, offering practical solutions. He saw issues more clearly and held a more rational, systematic perspective than Forrestal did. Eberstadt brought order out of chaos, particularly the type of intellectual disorder that marked Forrestal's accumulation of ideas ranging from welfare capitalism to archconservatism. Eberstadt tried to find a middle ground between the extremes. He thought that attachment to ideologies bred political-economic conflict and class struggle. He reminded Forrestal constantly that problems could be solved through the correct organization of basic productive and administrative units of the state. Eberstadt discovered (and introduced to Forrestal) a set of organizational arrangements to solve problems in the theory and practice of corporatism.[3]

Corporatism described the political-economic vision and method that Eberstadt and Forrestal applied to the building of American national security institutions between 1940 and 1949. To Eberstadt and Forrestal corporatism meant a specific set of political-economic assumptions, interorganizational arrangements, and intermediary structures employed by an enlightened corporate element cooperating with government to build a "New American State." Their version differed from the theories and forms that late nineteenth-century corporatists in France, Austria, and Germany adopted in order to provide order in an increasingly chaotic industrial society. The European form emphasized the division of society into a few clearly defined noncompetitive interest groups, arranged hierarchically and protected by the state.[4]

The United States lacked these basic ingredients for corporatization. There were none of the socioeconomic concentrations, industrial trade associations, or parliaments found in Europe. American society contained individual, fragmented, and competing interest groups. Americans held strong liberal, democratic, and laissez-faire capitalist traditions. However, technological innovations and industrialization convulsed late nineteenth-century American society as much as it did industrial European states, and the new situation demanded new political-economic organization for orderly growth and development. American corporatists borrowed, and greatly modified, ideas and forms from European corporatism to fit the American system, in which power was dispersed among many interests. Instead of replacing liberal institutions, party government, or free-market capitalism, American corporatists sought to share power through associations where private organizations cooperated with government to form loose combinations of interests. In this associative state, organizational experts coordinated, advised, and mediated between interests to make the loose network mesh into a corporatist system. Corporate elements and government cooperated to build interorganizational structures, design policy, create a public consensus, and guide legislative action. The entire system in this scheme depended upon policy integration and interest representation through connecting mechanisms run by expert institutional coordinators.[5]

The American experience with corporatism occurred during an organizational and technological revolution that led to the formation of large-scale corporate, economic, professional, and governmental structures. During the first forty years of the twentieth century, American corporatists periodically (although in a largely uncoordinated fashion) built quasi-official bodies and informal associations that brought business and government together in their ideal of an enlightened public-private partnership. Organizations such as the National Civic Federation, founded in 1900, promoted cooperation between employee and employer, and between business and government. A corporatist impulse directed the formation of the U.S. Chamber of Commerce. The first world war stimulated formation of trade associations, consolidation of interests, and closer business-government ties through the War Industries Board, National Defense Advisory Council, and other mobilization agencies.[6]

During the 1920s, Herbert Hoover promoted voluntary political-economic associations at home and informal ententes abroad. The more liberal reformist elements in the business community fostered enlightened partnerships between business and government. Their influence resulted in the creation of the liberal-corporate Business Advisory Council that resided in, but received no official status from, the Commerce Department. During the Depression corporate liberals began to adopt elements of Keynesian economics by advocating that government spending programs and fiscal policy be used to discipline interest groups. The American experiment with corporatism reached a peak before the second world war with creation of the ill-fated National Recovery Administration (NRA). The NRA tried to eliminate costly competition by promoting trade associations and business-labor collaboration.[7]

Corporatism had not taken root anywhere in American society by the late 1930s, and, according to historian Kim McQuaid, it lay in stasis between 1935 and 1939. International crises and American rearmament created conditions for its revival. War planning, mobilization, military-industrial combinations, and extensive business-government cooperation provided essential steps toward corporatism, which brought organizational experts like Eberstadt and Forrestal to Washington to revive the languishing corporatist vision. Eberstadt recognized that this world crisis gave organizational experts an opportunity to build interorganizational structures and influence larger political-economic developments. "Business and labor, as we know it," he told Henry Luce, publisher of *Time* magazine, "now have their chance—and I think it's the last one—to emerge and do a job." Forrestal also thought that the crisis could be used to build a "practical nexus" between liberals and conservatives, labor and employer, and business and government.[8]

Eberstadt and Forrestal brought this corporatist vision with them to government in 1940. Though they held different levels of sophistication, understanding, and commitment to corporatism, they shared a political-economic perspective and set of principles that distinguished them as American corporatists in the mold of Bernard Baruch, William G. McAdoo, Owen D. Young, or Herbert Hoover. Fundamentally, Eberstadt and Forrestal agreed that undisciplined party politics and competitive economic interests engendered social conflict. Under the pluralistic system, Eberstadt observed, "Group has been aligned against group, class against class, and race against race." To eliminate some of this friction, Eberstadt and Forrestal advocated the organization of noncompetitive, self-governing, officially sanctioned economic units. Forrestal wanted all radio and cable services merged into one great private communications syndicate with limited government supervision to ensure enlightened operations. Eberstadt proposed a federally licensed, privately owned national airline syndicate: "It might be that a middle course consisting of non-competitive routes given to different companies would avoid the economic loss of undue competition, and at the same time furnish, in a measure at least, the benefits of competition and comparison of service and equipment."[9]

In such arrangements government served as arbiter between conflicting interests. The state intervened or regulated only to ensure a harmonious balance of interests. Government cooperated with corporate elements through commissions, advisory boards, and associations to develop political-economic policy. "Government alone cannot do the job," Forrestal explained, "and business cannot do it unless it has the full-out support of government both inside and outside, and by that is meant the lifting of as many restraints and time-consuming irritations as possible so that businessmen have time to devote themselves to the real problems."[10]

The Eberstadt-Forrestal brand of corporatism stressed leadership by an organizational elite that mediated between conflicting groups, generated public consensus, designed policies, and managed national affairs. The impulse for this concept came, at least partly, from their friend Walter Lippmann, a popular journalist and author. Lippmann advanced the idea that elite business administrators and industrial statesmen had to guide the modern American state through the chaos that followed industrialization. Eberstadt and Forrestal selected Lippmann as a potential investment partner for the proposed airline syndicate.[11]

Like Lippmann, Eberstadt and Forrestal believed that the nation's progress and prosperity lay in the hands of expert managers and organizers. Forrestal wanted "individual men with imagination, initiative and tact—and driven not by the profit and loss statement but by the desire to see government perform its function." Eberstadt specified national leadership by "conscientious high class men" in law, industry, business, education, government, and the professions and stressed the importance of technocrats who advocated scientific control of national resources and reorganization of society. He concluded that the United States had to be reorganized by "competent men of good will desirous of getting along together in the national interest," and he kept the names of such men on an elaborate inventory that he titled the "Good Man List."[12]

The Good Man List compiled by Eberstadt over four decades (and shared regularly with Forrestal between 1929 and 1949) led to the concept of the Good Man employed in this book. The list introduced the larger organizational sector and the milieu of Eberstadt and Forrestal. The list revealed a set of socioeconomic values, networks, and relationships that political scientist Philip H. Burch, Jr., classified as the basic ingredients for an American elite. A profile of the Good Man, taken from more than forty men who deeply influenced Eberstadt and Forrestal between 1929 and 1949 helped define the Good Man concept and its application to this study of national security organization. Sixty percent of the men from Eberstadt's inventory attended Princeton, Harvard, Yale, and Columbia universities or law schools. More than half represented an older Anglo-Saxon, Protestant establishment, but the list also contained sizable German Jewish and Irish Catholic American subgroups. Political affiliation seemed relatively unimportant for inclusion on the list, as there were comparable numbers of Republicans and Democrats and even a small independent group represented.[13]

The occupational profile of the men on Eberstadt's list revealed more about the Good Man concept. Nearly 75 percent of the men belonged to Wall Street law, investment, and banking firms. The remaining 25 percent held top management positions in the largest electrical, aviation, chemical, and communications companies. Significantly, the lawyers and bankers served these same industries as corporate organizers, planners, advisors, and publicists. More than seventy percent of the men on the list sat on advisory councils, quasi-official commissions, and governmental agencies during and between the wars intent upon developing the corporatist partnership of business and government.

The names on Eberstadt's list were there because the men possessed special talents in reorganizing or refinancing business and government and because they shared the same corporatist vision. Eberstadt later added the names of engineers and scientists because these experts assisted in what he saw as "the struggle of humanity for improvement of its physical, mental, and spiritual condition." In a sense, those on his list reflected the corporate, pro-capitalist reformism or liberal corporatism identified by historians Ellis W. Hawley, Thomas McCormick, Joan Hoff Wilson, and other scholars of an American brand of corporatism. Moreover, most of those in the Good Man composite have been identified as corporatists in studies by Jordan Schwarz, Kim McQuaid, Robert M. Collins, Michael Hogan, and Richard Werking.[14]

Seventy percent of those on the Good Man list went to Washington as so-called dollar-a-year men during the second world war. They held prominent positions in the Office of Strategic Services, Army Navy Munitions and War Production boards, and other mobilization agencies. They dominated the civilian leadership posts in the War and Navy departments. As these Good Men organized the American war machine, they reintroduced the corporatist assumptions and structures that had been in stasis since the mid-1930s. Their path toward corporatism lay in the building of national security institutions that culminated in the National Security Act of 1947 and the Marshall Plan for European economic recovery.[15]

National security organization meant the orderly arrangement and linking of the different (and often competing) economic, social, and political elements of the state. "National security policy is like a mosaic made up of a multitude of actions and relationships," Eberstadt explained, "but, like a mosaic, there must be a guiding and dominating theme, and in our case that theme should be the maintenance of peace by all means and forces at our disposal." These forces included a strong defense and consistent foreign policy tied to national resources. National security required a unified and informed democratic society with a stable domestic economy in which labor, management, industry, business, and government cooperated to achieve maximum production with minimal social conflict or wasteful competition.[16]

As they labored to define and build national security institutions between 1940 and 1949, Eberstadt, Forrestal, and the Good Men continued a process begun by (and at times received help from) an earlier generation of corporatist national security organizers. That earlier generation had searched between 1890 and 1919 for mechanisms to rebuild the American political economy, which had been disrupted by industrialization and international competition. These pioneer "New American State" builders idealized corporate business practice and scientific management, with its emphasis on efficiency. They searched for ways to relate economic and military organization to national policy. Elihu Root and others, who historian Richard Werking called the master architects of the Theodore Roosevelt and William H. Taft administrations, began to reorganize the War, Navy, and State departments. They organized the Joint Army and Navy Board, Navy General Board, and Army War College but failed to establish a national defense council or central budgeting office.[17]

The Woodrow Wilson administration continued to reorganize the machinery of state. William Redfield, secretary of commerce and labor, and William G. McAdoo, secretary of the treasury, recast their executive agencies to promote business efficiency at home and expansion of American economic power overseas. During the first world war, Wilson linked industrial and military policy in a primitive national security partnership. The Selective Service System, the Naval Consulting Board, National Research Council, and Council of National Defense and its Advisory Commission connected business and government on an unprecedented scale. When these arrangements "for the national security and welfare" proved inadequate, Congress gave Wilson authority under the Overman Act of 1918 to coordinate through executive reorganization.[18]

In the end, Wilson established temporary agencies for national security. One of these was the War Industries Board (WIB), headed by Bernard Baruch. He and other Wilson appointees relied on voluntarism, cooperation, and private association rather than formal organization to accomplish their objectives. They assembled experts and sought to organize industrial trade associations. The war ended before the WIB or other emergency machinery solidified into permanent institutions. Fragments such as a statistical office found their way into regular Commerce Department channels, while the Council of National Defense stayed on the statute books as an inactive agency.[19]

National security organization received little coordination during the interwar years. The National Defense Act of 1920 provided for the Army Navy Munitions Board (ANMB) in the office of an assistant secretary of war. A central budgeting agency established under the Budget and Accounting Act of 1922 held implications for future national security. In the meantime, W. F. Willoughby of the Brookings Institution and other public administration scholars developed theories and blueprints for a modern national security state. Mostly, national security organization lay in advisory commissions like the group of corporate experts assembled in 1925 to study aviation and national defense organization for President Coolidge. Later, Forrestal idealized this commission, led by Wall Street banker Dwight Morrow, as the perfect format for national security organization for the post–World War II era.[20]

Baruch and WIB veterans kept their wartime mobilization experiences alive through the promotion of a "cooperative industrial democracy organized for war." They lectured on this concept at the Army War College. They educated public and government throughout the interwar period on the relationship between the political economy and national security. They advanced plans for noncompetitive industrial committees, national resources inventories, and labor-management cooperation. Baruch's organizational experts taught that economic planning prepared the United States in peacetime to fight a modern world war. They emphasized that the United States had made a serious mistake in dismantling the WIB, rejecting the League of Nations, and neglecting national defense organization after the Great War. The work begun by Root and others and advanced during the world war remained incomplete. Now on the eve of another world war, the United States still lacked basic structures for national security, forcing the government to turn to experts from the private sector such as Eberstadt and Forrestal to organize war preparations.[21]

Between 1940 and 1949 Eberstadt and Forrestal gradually assembled pieces of a national security mosaic. During this period they found themselves in positions to influence directly the building of the entire modern national security establishment. Their public-private partnership and their association with those on Eberstadt's Good Man List allowed the two investment bankers to embody corporatist values and forms in actual organizations. In order to better understand these values and forms, and how Eberstadt and Forrestal developed a corporatist perspective, the following chapter traces their education and experiences with the organizational revolution that shaped the new American state between 1909 and 1939.

2. The Partners' Education, 1909–1939

The intensity with which Eberstadt and Forrestal pursued national security organization between 1940 and 1949 arose from thirty years of observing and actively participating in the organizational developments that transformed the United States. Their response to crisis, the values they carried to government, and the structures they helped to build arose in large measure from their shared education. That education extended from Princeton University to the Plattsburg military training camp movement, from Wall Street to Washington, and through Good Man networks and associations. In part, these explained the corporatist organizational values and forms that Eberstadt and Forrestal grafted upon the American national security state.

The partners' education began when Eberstadt entered Princeton University in 1909. Woodrow Wilson was president of Princeton and dominated the campus. He urged Eberstadt and other incoming freshmen to attend Princeton as preparation for service to the nation. At the moment, Wilson was an embattled administrator, fighting to eliminate exclusive upperclassmen and faculty eating clubs. He developed a tougher graduate program against entrenched opposition. All the while, Wilson struggled with his own philosophical transformation from inherent conservatism to progressive reformism. Eberstadt confronted the combative Wilson a few weeks into his first semester, when along with seven classmates he was suspended by the college president for "roughhousing." Readmitted a month later, the chastised freshmen received Wilson's lecture on duty: "They are to regard themselves as under special obligations, when they return, to behave themselves with order and propriety."[1]

Order and propriety formed part of Eberstadt's German American background. Order and propriety had also been the code at the elite Newark Academy where he ran track and graduated with honors in 1909. Discipline at this private northern New Jersey prep school, Eberstadt recalled, was maintained by a firm paddling "via the seat of the pants." Eberstadt thrived at Princeton University. His Venezuelan-born mother, Elenita

Lembcke, had introduced Eberstadt to music, literature, and the classics. His father, Edward Frederick Eberstadt, had studied at Heidelberg University and was a mean-tempered New York import-export merchant with A. D. Straus and Company. He tutored his son in German culture and political-economic theory, and the younger Eberstadt received top grades at Princeton, graduating Phi Beta Kappa in 1913.

Eberstadt was a campus leader, handsome and dark-eyed, expensively dressed and immaculately groomed, and he fit in perfectly with Princeton's clubby social life, aggressive sport, and progressive academic program. He served as freshman class president, on the Bric-a-Brac and dining hall committees, and on the student board of athletic control. He joined the Literary Club, wrestled, and managed the football team. Like Woodrow Wilson three decades before, Eberstadt became editor of the *Daily Princetonian,* the campus newspaper. This experience turned out to be most important, for it introduced him to James Forrestal.[2]

Eberstadt first met the slightly built, intense Forrestal in 1912. "I remember well early in the fall of 1911 [*sic*], I think it was, a freshman, wearing the conventional black jersey and black skull cap, joined the group who were heeling for a job as one of the editors of the *Daily Princetonian,*" Eberstadt recalled. "The individual in question was a serious, quiet–one might say obscure–person." Forrestal was the son of an Irish-immigrant construction contractor and a Catholic school teacher. Leaving his small hometown in upstate New York, Forrestal first entered Dartmouth College but transferred to Princeton. Forrestal's father was a local Democratic Party boss and postmaster who had supported Grover Cleveland, a Princeton resident until his death in 1908, and currently promoted the political career of the young Franklin Roosevelt. The younger Forrestal had become a political reporter for a Poughkeepsie newspaper and covered Franklin Roosevelt's local campaigns in 1910. Undoubtedly he considered Princeton a better place than Hanover, New Hampshire, to launch a journalistic career, for at Princeton, he could be reporting the activities of Woodrow Wilson, elected governor of New Jersey in 1910. Ambitious and driven to succeed, Forrestal determined to become the *Princetonian*'s best reporter. "Before very long," Eberstadt noted, "the other heelers discovered that often when they sought articles, the material had already been promised to Forrestal."[3]

The *Princetonian* connection initiated an unlikely relationship between two students of markedly different personalities and backgrounds. The college friendship, itself, was understandable. Forrestal proved himself a top reporter on Eberstadt's staff, and both shared enthusiasm for campus athletic programs and club life. Forrestal was bright and witty and could be charming. He appeared eager to win friends among the more secure and wealthy Protestant students like Eberstadt. Soon, Forrestal and Eberstadt developed a subtle chemistry beyond their chummy college relationship, and it tied them together for the remainder of their lives. During the 1920s they developed a confidential lawyer-client relationship, became business partners, and traveled in the same New York society circles. They purchased property within a few miles of each other on Long Island and joined the same clubs and organizations.[4]

The Forrestal-Eberstadt friendship cannot be understood only in socio-economic terms, or as traditional male comradeship. They rarely fished, boated, or played golf or croquet together, as Eberstadt did with Allen Dulles, W. Averell Harriman, Herbert Bayard Swope, and those on his "Good Man Fishing List." Possibly the career-driven Forrestal—who practically lived at the office and for his work—simply did not have time for this type of relationship. Yet Eberstadt, polished, conservatively attired, and cigar-smoking, met high-strung Forrestal daily. Together they attended the Wall Street routine of conferences, luncheons, and dinners. All along, they gave each immediate and frequent access to the other at any time and no matter how busy the other might be. No one else shared a similar arrangement with either. Some colleagues and associates found both men impatient and unfriendly, but Forrestal and Eberstadt appeared completely at ease with each other and with the small circle on the Good Man List.

Forrestal called upon Eberstadt constantly for advice, counsel, and assistance. There was always a bit of that early editor-reporter (or lawyer-client) relationship between the two. Incessant discussion of business, government, and world affairs surrounded their friendship. "The relationship . . . was not 'intimate' in the conventional sense of that word," Eberstadt's son recalled, adding that "there seems to me a good possibility that they never discussed some personal matters that most men talk about with their close friends." Nevertheless, closest associates insisted that they remained devoted to each other and that Eberstadt exerted tremendous influence on Forrestal. This seemed particularly true between 1940 and 1949, when Eberstadt and Forrestal discussed organizational strategy almost daily. In their conversations, Eberstadt served as teacher and organizer. Others influenced Forrestal as well, but no one guided his public life and conduct more than Eberstadt did.[5]

The permanence of the Eberstadt-Forrestal connection arose from their college experience. "Such colleges as Princeton are not intended to turn out common men," Eberstadt insisted. This pervasive attitude welded lifelong bonds. Even reclusive undergraduate George Kennan, who later influenced Forrestal's thinking about the Soviet Union and Cold War containment policy, recognized that Princeton determined his future life and career. "I felt that I was fairly treated at its hands; I respected it intellectually: I took pride in it as an institution," Kennan recalled. Though he never graduated with his Class of 1915, Forrestal received help from Princeton friend Dean Mathey, who made certain that Forrestal found a good job with a New York City investment house. Forrestal extended similar assistance throughout his life to young Princeton graduates. "Many a boy now established in business can thank Forrestal for the sums which put him through college," Eberstadt told a reporter in 1947.[6]

Although Forrestal rejected other outside interests that drew him away from his work, he always found time to visit Princeton and to communicate with its president, Harold Dodds, and faculty members. In 1946, Forrestal and Eberstadt served as key members of the rededication ceremonies commemorating Princeton's two-hundredth anniversary. "No one was more loyal and interested in his Alma Mater," insisted John T. Connor, Forrestal's assistant and a future secretary of commerce. Indeed, the larger-than-life bust of Forrestal that now stares at researchers from the corner of Princeton's Seeley Mudd Library reading room seems to symbolize this deep commitment.[7]

Eberstadt's loyalty to Princeton caused him to defy the president of the United States. During World War II, Roosevelt called Eberstadt to the White House and asked him to fire Thomas Armstrong, a War Production Board (WPB) assistant, over an alleged conflict of interest between his Standard Oil Company connections and government contracts. Eberstadt refused, standing by a fellow member of the Class of 1913. This action probably contributed to Eberstadt's eventual dismissal from his wartime post as WPB vice-chairman. Other Princeton connections proved less dramatic but equally enduring. Campus friendship with Allen Dulles developed into a lifelong comradeship with Allen and his brother, John Foster Dulles. Common interest in the university tied Eberstadt to Arthur Krock, Bernard Baruch, Herbert Bayard Swope, Courtney Johnson, and David Lawrence, the last having been editor of the *Princetonian* when Eberstadt joined the staff in 1909.

Lawrence introduced Eberstadt to the press's role in educating the public and to his fervent Wilsonianism. Lawrence later was a conservative editor of the *United States News* and a critic of the New Deal, but in 1909–10 he admired Wilson, a progressive Democrat. Lawrence used the college newspaper to promote Wilson's candidacy for governor of New Jersey. Wilson's political-economic thought vied with sport and social activities for headlines in the campus newspaper. Wilson urged Eberstadt's generation to undertake the "one great task of adjustment" between business and government, and between labor and capital. He advocated moderate government regulation to restore the "symmetry" between individualism and order that industrialization had upset. Wilson insisted, "Our business is to adjust right to right, interest to interest, and to systematize right and convenience, individual rights and corporate privileges, upon the single basis of the general good."[8]

Wilsonian politics of adjustment influenced Eberstadt's entire career as corporate lawyer, investment banker, and government official. Wilson's administrative theories also made a permanent impression. Wilson crusaded between 1910 and 1912 for economy and efficiency in government, a theme common to progressive politicians who applied the era's pervasive theory of scientific management to government administration. In a speech before the New Jersey mayors' conference in 1912, Wilson called for compact organization, perfect coordination, and systematic administration. "We are entering upon a period of remarkable business efficiency all over the country, and it seems to me that we should make every effort to introduce this particular kind of efficiency into government." Thirty-five years later in a radio address, Eberstadt echoed Wilson as he explained his own quest for "ways to improve the efficiency, economy, and quality of the services to the country."[9]

Eberstadt, like Lawrence, later cooled toward Wilson because of the president's interest in more governmental regulation of the economy and society. Eberstadt praised Wilson's educational reforms and emphasis on public service and duty to the nation but also criticized Wilson as a politician "more disposed 'to make history' rather than follow it." Despite close contact with Wilsonian democracy, Eberstadt and most of his classmates became members of the Republican Party. Eberstadt's self-styled "liberal Republicanism" paralleled Forrestal's conservative Democratic Party principles, blurring distinct political lines. Eberstadt explained in a signed *Princetonian* editorial in March, 1912, that all men remained either conservatives or liberals. He lumped in the latter category Wilson, Robert M. LaFollette, and Wall Street lawyer Elihu Root. He placed only William H. Taft in the conservative camp. "I maintained stoutly that I had always been a liberal," Eberstadt wrote in 1943. In reality, as his 1912 college editorial revealed, political labels and ideologies meant less to Eberstadt than did the moderate interventionist and organizational legacy of the Progressive Era.[10]

Eberstadt received indoctrination in the organizational assumptions of the Progressive Era from Prof. Carl Albert Fetter. Fetter arrived on the Princeton campus to chair the Department of History, Politics, and Economics in time for Eberstadt's senior year. Already a leading welfare economist, president of the American Economic Association, and popular lecturer at Cornell, Fetter brought a new perspective to Princeton. He shared the approach of progressive economists Simon Patten, Richard Ely, and other reform-minded American scholars who had learned their craft under German social scientist Johannes Conrad at Halle University and Austrian economist Eugene von Böhm-Bawerk. Instead of teaching the classical theories that natural forces governed the laws of supply and demand, the German-trained economists taught that experts using historical and sociological data could reform those laws by working through the state to promote production. Through economic planning, government and private experts could manage the use of natural resources, replace scarcity of supply with abundance, and raise the standard of living.[11]

Fetter taught Eberstadt that moderate state intervention could result in a more efficient economic system. The state would protect industry, labor, and agricultural interests and tie them to larger national development. An efficient state sought a course between free-market enterprise and state regulation. "Formerly the line between the action of the state and that of private and ecclesiastical corporations was dim," Fetter explained. "Public functions were exercised by many guilds, societies, and church corporations, but the modern state has gradually taken over these functions," he wrote. Fetter participated in the movement for closer cooperation between private associations and public agencies. He studied the settlement house movement and joined the American social work movement that sought to develop public policy for improving the standard of living and avoiding social conflict. New York's progressive governor, Charles Evans Hughes, appointed Fetter as an advisor to the State Board of Charities. Fetter served a similar expert advisory function on the New Jersey Conference of Charities and Correction and on the New Jersey Conference for Social Welfare.[12]

Fetter's activist welfare economics and view of the state deeply influenced Eberstadt, who later applied many of Fetter's associative theories to his own political-economic blueprints for post–World War II reconstruction. Fetter also introduced Eberstadt to William C. Redfield's *New Industrial Day,* a book that contained a fundamental outline for Wilsonian commercial policy. Redfield, a New York congressman and industrialist, voiced the Democratic Party's position on tariff revision as the means to stimulate business, foreign trade, and international harmony. His theories impressed Wilson, who selected Redfield as his secretary of commerce, a role that Redfield used to promote private business expansion abroad. "Our foreign trade is a safety valve," Redfield announced, "that relieves the pressure of over-production at home."[13]

Eberstadt analyzed Redfield's views for Fetter's class. Redfield's work had taught him the need for the adjustment between labor and capital, and between political interests and duty to the nation. Redfield's impact on Eberstadt appeared more profound because he emphasized German commercial policy as the model for American business. German American merchants (like Eberstadt's father) used modern management techniques, cost accounting, and production forecasting. They fostered cooperation between labor and management to give them a competitive advantage. "[German merchants] are backed by the most perfect public and private organizations on earth by schools in which men are trained from their boyhood to the patriotic and lucrative purpose of expanding German trade," Redfield concluded.[14]

Eberstadt followed Fetter's path after graduation and studied in Germany. However, in 1913 Europe was not as it had been in the 1880s, when Fetter had visited Johannes Conrad at Halle, The older generation of social scientists was dying out. The intellectual freedom that had attracted American scholars existed no longer. Now, German social scientists served not Bismarck's social welfare state but an authoritarian German state preparing to fight a world war. It seemed to American critics that German scholars had become apologists for Prussian militarism. Eberstadt encountered this environment as he attended the universities of Berlin and Munich during 1913 and early 1914.[15]

Nevertheless, Eberstadt found stimulating teachers in labor historian Heinrich Herkner, political scientist Ludwig Bernhard, and his own uncle, Rudolph Eberstadt, a respected German social scientist at the University of Berlin. Wherever he studied, Eberstadt met proponents of the corporatist state. The aging Lujo Brentano lectured at Munich on the need for responsible industrial cartels and labor organizations to create social order through public-private committees that strove to temper conflict. Scholars in Berlin presented forms of corporatism that stressed the organization of society into functional groups rather than by class or individual. Instead of traditional politics with elected legislators concerned with their own interests, the corporatists advocated industrial-vocational parliaments to coordinate parts of the state and society for the national interest.[16]

Eberstadt sympathized with aspects of German corporatist organization and theory. This Prussian spirit, liberal American jurist Felix Frankfurter observed, was never far from Eberstadt. New Dealer Henry Wallace agreed but felt that Eberstadt's Prussianism escaped the fascist extremism implicit in some European corporatism. More to the point, Eberstadt's German education comprised part of a larger set of values developed between 1909 and 1914 that included the influence of Wilson, Fetter, and Redfield. If he borrowed from German corporatism, Eberstadt merged it into a larger and substantially different form of American corporatism. In fact, his German education discouraged Eberstadt. When he returned to the United States he found no one interested in publishing his work on the Public Employment Exchanges of Germany.[17]

Eberstadt considered a teaching career in the classics and then in 1914 enrolled at Columbia University's law school. Columbia presented an ideal environment for Eberstadt. Law professors George F. Canfield, Frank Goodnow, and Charles T. Terry had studied in German universities. Professor George W. Kirchwey, like Eberstadt, came from a prosperous German American family. Columbia's law faculty continued the Progressive organizational education already introduced to Eberstadt at Princeton by Fetter. Like Fetter, Kirchwey promoted philanthropy, penal reform, and social welfare organizations and helped to found the New York School of Social Work. Professor Terry's organizational experiences included membership on the New York Civic League and National Civic Federation, the latter embodying an American form of corporatism during the Progressive Era.

Columbia also introduced Eberstadt to a network that was connected to Harlan Fiske Stone, dean of Columbia's law school. These connections included Dwight Morrow, Jackson Eli Reynolds, and the Wall Street firm of Sullivan and Cromwell. In one way or another, all played central roles in Eberstadt's future. Reynolds became Eberstadt's law school mentor, helped place him on the law review, and later joined him as legal expert on Owen Young's Commission on Debts and Reparations of 1929. Most important, Morrow's friend and associate Joseph P. Cotton hired Eberstadt in 1916 for a summer internship with his Wall Street law firm of Spooner and Cotton, and after World War I Cotton brought Eberstadt into full association with McAdoo, Cotton, and Franklin.[18]

While Eberstadt excelled at Columbia law school, his friend Forrestal left Princeton a few weeks before graduation, possibly because of a feud with a teacher. After working briefly as a sales representative for a tobacco company, Forrestal joined Herbert Bayard Swope's *New York World* staff, coming in contact with New York cafe society that included his future wife Josephine Ogden, a *Vogue* magazine editor. Forrestal developed connections with the indefatigable Swope, Baruch, and their circle of Long Island friends and playmates. Then, Princeton graduate Dean Mathey hired Forrestal as Albany sales representative for the investment firm of William A. Read and Company, later renamed Dillon, Read. Soon, the *New York Times* called Forrestal "one of the best known bond authorities in the financial district." Forrestal's success, Eberstadt explained, lay in his ability "to choose good men" to sell bonds for him.[19]

Service in World War I interrupted Forrestal's burgeoning investment career and Eberstadt's law school classes but provided another stage in their education. Their contrasting wartime experiences suggested, too, the chemistry of their later partnership. Forrestal scrapped to get into the fighting. For some reason recruiters for both the army and marines rejected him. He enlisted finally in June, 1917, as a seaman second class in the U.S. Navy. Once again the Princeton connection determined Forrestal's course as he joined classmates in Bayshore, Long Island, where one student's mother financed flying lessons. In October, Forrestal trained with the U.S. Naval Aviation Detachment of the Royal Canadian Flying Corps and then with U.S. Navy aviators in Virginia. At least one crash nearly killed him. Forrestal awaited assignment, hoping for aerial combat in Europe such as that experienced by his friends Robert A. Lovett and future roommate Artemus Gates of the Yale Flying Unit. Instead, the navy ordered Forrestal to a Washington desk job in the recently created Office of Naval Operations. He served as a supply officer for the struggling naval aviation service, which was so short on aircraft that it could not provide flight duty for already-trained Forrestal.[20]

Meanwhile, Eberstadt volunteered, along with many of his Wall Street associates, in the Plattsburg training camp movement begun by Gen. Leonard Wood but now run by corporate lawyer Grenville Clark, of the firm of Root, Clark, Buckner and Howland. Plattsburg meant more than the preparation of civilians to command a citizen army in the event of war. "It means, first of all," General Wood explained, "the moral organization of the people, an organization which creates in the heart of every citizen a sense of his obligation for service to the nation in time of war or other difficulty." Plattsburg graduates had a duty to organize industrial and military resources to meet modern defense requirements. These "armed progressives" felt that the traditional American system of national defense remained costly and inefficient. The nation needed Plattsburg to inculcate a sense of national duty, order, and progressive values, since government remained unresponsive to national security. Thus, private experts had to prepare the nation for war.[21]

Plattsburg reinforced Eberstadt's emerging concept of the state's intermediary role and association with private experts. It also brought him into contact with many individuals of similar background and views who later joined Forrestal and Eberstadt to build a national security establishment. Through the Plattsburg movement and attachment to the 77th Division, Eberstadt met John J. McCloy, Lewis Douglas, William Draper, Robert P. Patterson, and Henry L. Stimson, who was Eberstadt's colonel during World War I.[22]

Bob Patterson became Eberstadt's closest friend and through him a Forrestal comrade as well. Patterson embodied Eberstadt's Good Man. Like Eberstadt, Patterson achieved Phi Beta Kappa, stood at the top of his law school class, and edited the law review. Both trained under the Cravath legal system whereby the best Harvard or Columbia law school graduate joined the top Wall Street firm and worked his way up to partner by mediating in conflicts between corporate clients. Through Plattsburg, Patterson and Eberstadt went to the Mexican border with the New York National Guard in 1916 and disembarked from Camp Upton for the European war zone. Both fought on the Western Front, where Patterson received wounds and Eberstadt suffered from gassing at the Vesle Defense Sector.[23]

Forrestal envied those who, like Eberstadt and Patterson, had combat experience and war wounds. His war service paled by comparison. Nevertheless, it shaped his attitudes toward national security as much as Plattsburg influenced Eberstadt and the others. During a seven-month stint in Washington, Forrestal developed a "Frank and Jim" relationship with his father's political ally, Franklin D. Roosevelt, now the assistant secretary of the Navy. Together, they advanced early development of naval aviation, and they had a mutual understanding of the problems of industrial mobilization. Though Forrestal held a minor post, one naval aviator insisted that "he was truly the key expediting officer for the requisitions etc. in this pioneering branch." Forrestal admired John Towers and other early naval aviators, becoming eventually an advocate of a Mahanian hybrid concept of sea-air power as the first line of national defense. Simultaneously, he shared Roosevelt's aversion to a separate air force, preferring to maintain naval aviation as an integral part of the fleet.[24]

After the war, Forrestal rejoined Dillon, Read and Company, while Eberstadt went to work for McAdoo, Cotton and Franklin. These employers also added to the shaping of Forrestal's and Eberstadt's political-economic and organizational values. Forrestal's boss, Clarence Dillon, had been one of Baruch's assistants on the War Industries Board and during the next decade worked with him to develop a "New Era" for American business. Dillon associated with other war mobilizers and New Era organizers, including Paul Cravath, Owen D. Young, John Hancock, Harrison Williams, H. B. Swope and his son Richard, Herbert Hoover, and Robert S. Lovett (Robert A. Lovett's father). Naturally, Forrestal moved into this Wall Street crowd. By 1923 he had become Dillon's partner. The extensive banking, business, and society connections that Forrestal developed helped form the attitudes and organizational methods that he would apply to the construction of national security machinery.[25]

Eberstadt also joined Wilsonian war mobilizers. William Gibbs McAdoo, his first boss after the war, was Wilson's son-in-law and a former secretary of the treasury. McAdoo envisaged his law firm as an intermediary between private enterprise and government policy. He worked to advance (with government cooperation) Anglo-American telegraph and telephone companies and multinational petroleum syndicates. More significantly, McAdoo reinforced Eberstadt's growing conviction that industrial America had to pursue a course between state control and destructive free enterprise, and between business and government leadership. McAdoo also taught the importance of mobilizing public opinion to support the orderly organization of American society. McAdoo left an impression on Eberstadt. Eberstadt wrote that McAdoo was the first "'great' man" of the many Eberstadt met during his lifetime. At the same time, McAdoo predicted a brilliant future for Eberstadt based on his tireless efforts for the firm's corporate clients.[26]

McAdoo's partner, Joseph P. Cotton, also played a central role in Eberstadt's education, contributing to Eberstadt's emerging Good Man concept. A brilliant Harvard law graduate, law review editor, former Cravath partner, and skilled negotiator in corporate reorganization, Cotton exemplified the progressive corporate lawyer in the mold of an Elihu Root or Henry Stimson. More liberal than Root or Stimson, Cotton earned a reputation on Wall Street for his defense of organized labor, social welfare groups, and citizen commissions for good government. He joined Paul Cravath in support of the settlement house movement to improve the standard of living and to curb social unrest. He sought to control monopoly and regulate industry for its own benefit. "The object of this reform is not to humble or cramp corporate business," Cotton insisted, but rather to render it more responsive to national interests. The U.S. Constitution gave the state the power to regulate so long as "it stops short of folly." Moderate government intervention, he continued, "shall not too far awaken the fears and sense of insecurity which the Tories feel in each change of the traditions of government or each new governmental interference in industrial matters and the relation of labor and capital."[27]

Eberstadt learned from Cotton's quasi-corporatist (or associational) approach to government. Cotton represented a number of organizations that linked public and private interests. His clients included the New York State Commission on Workmen's Compensation, Citizen Commission on Vice, and the State Federal Reserve Board, for which he counseled the right to regulate private banks, establishing, according to Cravath partner William Swaine, the constitutionality of the entire Federal Reserve system. In the Food Administration, under Hoover, Cotton directed the Meat Division during World War I, employing it as the arbiter between private producers and public consumer. Cotton used his licensing and contracting authority to balance private supplies and government purchases with a minimum of federal coercion or regulation. Next, Cotton served in London as an expert on the International Food Mission and Interallied Finance Council. As a member of these international bodies of private experts, Cotton noted the opportunities for U.S. bankers and businessmen to work with their foreign counterparts to fix prices, establish production schedules, and control supply and demand. Together, they could rehabilitate war-torn European economies, stabilize currencies, and promote orderly economic growth without the class conflict that he believed promoted socialism.[28]

Cotton explained his model for economic association in a 1919 article on the Wheat Executive that he prepared with Morgan banker Dwight Morrow. This Wheat Executive, a body of Allied representatives who controlled food supply and distribution, provided what Cotton and Morrow called the "common counsel." Instead of individual competing nationalistic agencies, the Wheat Executive encouraged international cooperation. It stabilized the world economy and raised the standard of living. Swaine believed that, with the Wheat Executive, Cotton had forecast the Marshall Plan for European economic recovery of the post–World War II era. More directly, the "common counsel" mechanism became central to Eberstadt's own Controlled Materials Plan during World War II.[29]

Eberstadt's daily work for McAdoo, Cotton and Franklin reflected the firm's organizational methods. He assisted the partners in corporate reorganization to avoid receivership during the post–World War I depression and then in mergers and expansion of foreign investment in the recovery that followed. Eberstadt's clients included RCA, General Electric, Goodyear Tire and Rubber, Chrysler and Dodge Motors, United Artists, and Allied Chemical and Dye, all major companies in the growth industries of the early 1920s: radio, automobiles, chemicals, motion pictures, and electrical materials. In each corporate reorganization or merger, he sought settlement with opposing but like-minded attorneys (those that he called the inside lawyers on his Good Man List). Eberstadt included them on his Good Man List because they all shared the goals of keeping clients out of bankruptcy court, refinancing firms through banking and investment houses such as Forrestal's company, and establishing a stable "New Era" corporate environment (or what historian Michael Hogan calls corporatist neo-capitalism). From time to time, they traveled to Albany or Washington to represent their clients in cases involving tax laws and regulations. They facilitated foreign transactions. They moved between the private and public sectors, adjusting, arbitrating, and reorganizing.[30]

Eberstadt's legal career between 1919 and 1924 reestablished connections with Forrestal that had been interrupted by war. Eberstadt's law firm represented Dillon, Read, and he became Forrestal's personal counsel, drawing up his will and taking care of other private legal matters. Eberstadt's initial visit to Forrestal at Dillon, Read revealed much about his hard-driving friend. He found Forrestal seated in the corner of a tiny ten-by-fifteen-foot cubicle that held only a desk and chair. Forrestal discouraged visitors. "This arrangement had the advantage of assuring him brief interviews," Eberstadt explained, "which fitted in with the rapid pace of his work." This suited the busy Eberstadt, too, as he provided the legal advice that enabled Forrestal to conclude the great Chrysler-Dodge merger, one of the largest transactions of the decade and one financed by Dillon, Read. Eberstadt counseled Forrestal on all such business, and advised "Baron" Clarence Dillon as well. Eberstadt's legal counsel propelled the firm to a number-three ranking among Wall Street investment houses.[31]

In 1924 Eberstadt prepared the Dawes Investment Brochure for Dillon, Read, which eventually opened the way for Eberstadt to join the firm as a full partner. This opportunity began to develop when a commission headed by Charles Dawes, Chicago banker and former budget director, and Owen Young, a General Electric executive, developed a plan to balance payments by Germans and debts owed to the United States by its allies. Unofficially the Dawes Commission represented American investors (and indirectly, Secretary of Commerce Hoover), who blamed reparations imposed on Germany for the unstable, inflationary economy in the potentially lucrative Central European market. They discussed a stabilization plan with Warburg banking, Thyssen industrial, and other German capitalist interests. They received helpful information on European conditions and a sympathetic hearing from Hoover, but no direct, active U.S. government commitment to adjust reparations and debts owed to the United States by Allied debtor nations. Finally, in January, 1924, the Dawes group had negotiated a modified reparations payment schedule with European representatives.[32]

The central adjusting mechanism in the Dawes Plan lay in a Transfer Committee of economic experts headed by S. Parker Gilbert, a former undersecretary of the treasury and Cravath partner. This committee determined Germany's ability to pay reparations and then balanced this with Allied demands for payments. The Transfer Committee encouraged investment in German industrial growth and sought to restore a vigorous new European capitalist society. The Dawes Plan was the classic expression of private-public cooperation and the corporatist neo-capitalist organization that punctuated New Era economics. At a testimonial dinner for committee member Young attended by Clarence Dillon and other New York financial leaders, Hoover praised the plan as a voluntary, unselfish solution to economic crisis and a peace mission without parallel in world history. Young claimed that the Dawes Plan initiated a "new day, a new vision, a new order."[33]

Several years after the Great Depression had dampened this buoyant attitude, Eberstadt reflected that "there was a kind of mental expediency in the New Era talk prevalent at that time." He admitted that "this disease had even infected what was then regarded as the Wall Street leadership." Certainly the New Era spirit persuaded Dillon to pursue European investments now that the Dawes Plan had stabilized affairs. Eberstadt's Dawes Investment Brochure explained how Dillon, Read might use the Dawes arrangements for German investments, and Dillon pushed his firm into the uncertain German market. However, his European representative failed to win the confidence of the most important businessmen. In response, Dillon (encouraged by Forrestal) offered Eberstadt a partnership if he would take over Dillon, Read's European investments.[34]

Eberstadt hesitated to join Dillon, Read. Though the job promised tremendous wealth in the wildly speculative atmosphere of the mid-1920s, Eberstadt enjoyed his attachment to Cotton and Franklin. He had become, in the view of clients, one of the best corporate lawyers and negotiators on Wall Street. Nevertheless, Cotton urged his protégé to take the leap. Besides, the investment business reunited him with college chum Forrestal. Indeed, Dillon, Read introduced Eberstadt to a new network of people in banking, insurance, and public utilities to place on the Good Man List. Dillon allies included the Rockefellers, Harrimans, Whitneys, and Aldriches. Eberstadt developed particularly close ties with W. Averell and Edward H. Harriman, working with them between 1925 and 1932 on a national airline syndicate and on multinational investment projects for Central and Eastern Europe.[35]

Eberstadt entered Dillon, Read at the time of its greatest expansion. Unrestrained by federal securities regulations, the firm used pyramiding, inside trading, and holding companies such as the U.S. and Foreign Securities Corporation to amass a fortune. Forrestal avoided taxes by placing profits in holding companies chartered in Delaware and Canada. Eberstadt did, however, escape the speculative mania or unethical dealings that would bring down the financial empires of such friends as utilities magnate Harrison Williams. Personally opposed to uncertain and greedy investments, Eberstadt followed his self-defined Wall Street ethic based on thorough investigation of prospects, their financial soundness, and the potential profitability of any investment. He employed Edward F. Willett, a young Princeton economist, to gather information on every aspect of a business deal. Eberstadt appeared less a Wall Street speculator than a modern financial expert and money manager for his clients.[36]

Eberstadt's service in Europe as Dillon, Read's representative kept him clear of the more tawdry side of the investment business. From his office at 39 Rue Cambon in Paris, Eberstadt contacted key industrial and banking leaders who were intent upon rebuilding economies and, in the words of historian Charles S. Maier, recasting bourgeois Europe. His network included what might be seen as an international version of the Good Man List: Dannie Heineman, Jean Monnet, Fritz Thyssen, Haljmar Schacht, Gustav Stresemann, and Warburg banker Carl Melchior. Melchior and the others stressed stabilization of the debts-reparations cycle, the infusion of American money to rebuild German industry, and the expansion of Polish and Romanian investments. Influenced by discussion with W. A. Harriman, Eberstadt recommended that Dillon, Read explore the Yugoslav and Soviet markets as well. "I have always felt that so far as the development of Russia was concerned," he wrote Forrestal in 1927, "it would be done principally by Germans with American capital." Eberstadt also urged his partner to work closely with Hoover and the Commerce Department in the development of international business.[37]

Dillon, Read concentrated on the German market. The Wall Street firm floated bonds for German municipalities, Catholic organizations, and industrial clients that included I. G. Farben and the Stahlwerke. Eberstadt coordinated the latter investment syndicate and reorganized five German steel companies with American money. "Our biggest success was the emission of the German Steel Trust bond," Eberstadt told the *New York Times* in August, 1927. Indeed, he had become an expert on German investment and was frequently interviewed and quoted. Eberstadt assured potential American investors that German bonds were a sound enterprise. The Dawes Transfer Committee had placed a control on the German economy. Eberstadt wanted German and American bankers and statesmen to cooperate under the Dawes formula to prevent destructive nationalistic economic policies.[38]

Apparently, Eberstadt sensed that selfish interests and unrestrained speculation threatened economic stability because in early 1929 he sold his partnership in Dillon, Read and Company. Stanton Griffis, a close friend and fellow investment banker, believed that Eberstadt left because the firm strangled three of its most ambitious members: Dillon, Forrestal, and Eberstadt. "They were too strong to stay together in one room," Griffis observed. Apparently Eberstadt wanted a larger percentage of the company. More likely, Dillon, Read's affinity for risky investments in faltering airlines, badly managed theater and hotel chains, and a disorganized printing ink business soured Eberstadt. Dillon's poor investments in Poland and Latin America bothered him as well. In any event, the sale of his partnership on the eve of the Great Depression netted Eberstadt between two million and three million dollars.[39]

Detachment from the investment firm freed Eberstadt to join Owen Young, David Sarnoff, and Thomas Lamont on the Young Commission on Debts and Reparations. Eberstadt viewed this group of experts as that common counsel mechanism projected by Cotton and Morrow in 1919. Eberstadt's move offered another opportunity to refine the Transfer Commission of the Dawes Plan. Eberstadt resumed earlier contact with Melchior, Monnet, and other European financiers. Together, they sought once more to stabilize the reparations-debt cycle through fixed terms. They tried again to build a moderate socioeconomic system that avoided the extremes of either fascism or socialism. Instead, Eberstadt discovered that political and economic self-interest choked the Young Commission debts conference.[40]

Eberstadt's own unhappy experience with Depression-era finance had him convinced that only a vigorous government-business partnership could deal with economic crisis. In 1930, he became entangled in the single bad business venture of his career as East Coast representative of the fragile Cyrus Eaton financial empire. After some reorganizing and consolidation, Eberstadt sold out at a loss. Next, National City Bank hired him to collect debts owed by Swedish associates of the bankrupt Ivar Kreuger Match empire that had fallen apart with the suicide of the notorious "Match King" in 1931. Concluding that mission, Eberstadt joined close friend J. Foster Dulles of Sullivan and Cromwell to reorganize the debt-ridden Hamburg-American Shipping Line. These tasks underscored in Eberstadt's mind the urgency of building cooperative organizations to prevent financial chaos.[41]

Eberstadt established a personal organization for sound investments when he formed F. Eberstadt and Company in 1931. He chose investors carefully. Averell and Roland Harriman and other trusted names from the Good Man List provided working capital. He brought Edward Willett into the office as an expert economist. He drew up a "Good Man Syndicate Investors List" that included William J. Donovan, David Bruce, Wendell Willkie, Marshall Field, H. B. Swope, and Averell and Roland Harriman. He asked Forrestal to become his partner. The vision of Eberstadt, Forrestal and Company as a leading Wall Street investment house disappeared, though, when Dillon offered Forrestal the presidency of his firm. The ambitious Forrestal accepted Dillon's more lucrative proposal. Alone, Eberstadt expanded his fledgling company. He concentrated on the reorganization of small- and medium-sized companies that were bankrupt and short on capital but that possessed good organization and growth potential.[42]

While Eberstadt launched his business, friends at Dillon, Read became the focus of a U.S. Senate Banking Committee investigation under New York prosecutor Ferdinand Pecora. The committee blamed the speculative and unethical practices of firms like Dillon, Read for the stock market crash and Great Depression. The Pecora investigation called Forrestal to testify in early 1933. Eberstadt counseled his former business partner. He compiled with David Lawrence, a Washington newspaper correspondent, a list of directions for Forrestal to follow when testifying before the committee. They advised Forrestal to have dress rehearsals and to answer all questions honestly. Following their advice, Forrestal explained candidly that he had avoided paying taxes during the 1920s through his investment trusts and holding companies.[43]

The Pecora investigation pushed Eberstadt and Forrestal toward support of early New Deal measures to regulate security and exchange transactions, the stock market, and investment banking. Forrestal's support for Roosevelt's program was not unexpected, given his political, personal, and family connections with the new president. However, Eberstadt's approval was surprising. A lifelong Republican and great admirer of Hoover, Eberstadt nevertheless endorsed Roosevelt's banking and stock market reforms and temporary emergency unemployment relief measures. He favored moderate government intervention to stem inflation and prevent speculation. He welcomed Roosevelt's selection of a McAdoo ally, Dan Roper, for commerce secretary, and Roper's support for a Business Advisory and Planning

Council (later the Business Advisory Council [BAC]) that included Young, Swope, and W. A. Harriman. Eberstadt endorsed the National Industrial Recovery Administration and helped Harriman recruit executives and lawyers for this quasi-corporatist New Deal agency that sought to link business and government. Eberstadt gave Harriman names from the Good Man List, including those of William Chamberlain, United Light and Power Company chairman; George S. Franklin, a former law partner; businessman Floyd Odlum; Paul Mazur, of Lehman Brothers; and Wendell Willkie, utilities counsel. "They are all personally known to me or to somebody else in this shop," he assured Averell Harriman.[44]

Eberstadt's main connection to the New Deal remained Roosevelt's advisor, Thomas Corcoran, "one of my boys," Eberstadt recalled. Corcoran had come out of Harvard law school in 1927 and joined Eberstadt's old law firm. He had also served for a time as counsel for Dillon, Read. Corcoran became friends with both Eberstadt and Forrestal, and when he joined the New Dealers in Washington, he continued to consult with his Wall Street comrades on securities, banking, and public utilities reforms. These comrades sent information to Corcoran (and to William O. Douglas and Joseph Kennedy) on how they might close tax loopholes, regulate the market, and promote business-government cooperation. Later, Corcoran reminded FDR that Forrestal had helped the administration develop the Securities and Exchange Commission.[45]

Eberstadt's initial support for the New Deal faded. He knew that the Depression's severity required direct federal action but believed that permanent recovery lay in government stimulation of normal corporate business activities. He wanted taxes on excess profits reduced, trade barriers removed, and business deregulated. Instead, the New Deal moved toward more regulation, more spending, and more protectionist trade measures. Inflationary spending forced Eberstadt's friend Lewis Douglas to resign as Roosevelt's first budget director. The NRA faltered and collapsed. FDR ignored BAC advice. The London Economic Conference failed to stabilize foreign commerce or currency, and Eberstadt blamed Roosevelt for sabotaging the economic conference in what he called the gravest act of American isolationism. Eberstadt concluded that "the burdens and handicaps which the New Deal had inflicted on this country will take years to repair."[46]

Recovery came less from New Deal policies than from the general mobilization of American defense industries during the late 1930s. Eberstadt recognized that deteriorating world conditions and the rise of the European dictators forecast another world war. He started financing machine tool and chemical industries, which were sure to prosper with rearmament. However, he opposed American military involvement in the war that exploded in Europe in September, 1939. "I have always regarded war as a destroyer of both wealth and confidence," he explained to journalist Arthur Krock. Possibly influenced by close business and personal ties to Germany, Eberstadt shared some of the isolationist sentiments of two intimates, William Chamberlain and investment banker Paul Shields. Nevertheless, he saw in the world crisis an opportunity to reorganize American production and to coordinate labor, agriculture, and business along the lines suggested by Hoover, Young, and other New Era corporatists. He agreed with corporation lawyer William O. Douglas that it was time to "build a structure of democracy based on security and stability" and to recognize that the "delicately interdependent society" required intermediary mechanisms for an active business-government partnership.[47]

Eberstadt joined critics of the New Deal who thought that Roosevelt seemed oblivious to such things. He had no blueprint, no program to reorganize America or to mobilize economic resources to confront totalitarianism. He neither supported nor opposed legislation that reduced corporate taxes. He ignored advice from Baruch for the establishment of an industrial mobilization agency. Pressed by Louis Johnson, FDR's interventionist assistant secretary of war, the president authorized a War Resources Board manned by those who fit Eberstadt's Good Man profile, such as Edward Stettinius of U.S. Steel, Harold Moulton of the Brookings Institution, and Lehman Brothers banker John Hancock. Three months later, FDR abandoned the board, hoping to avoid what one New Dealer called "the road to full, permanent cartelization of industry." Roosevelt also refused to employ the Army Navy Munitions Board as a coordinating body.[48]

Eberstadt began to think that Wall Street neighbor Tom Dewey could defeat Roosevelt in 1940. "Dewey's chances are excellent," he wrote William Chamberlain in February. Eberstadt reviewed speeches for the New York governor and planned a campaign strategy session with Chamberlain and John Foster Dulles. Dewey's speeches in early 1940 reflected their influence as he stressed the need for national unity and security. "National security," Dewey announced, "is based on national economic strength." Chamberlain and Eberstadt also advised Dewey to attack FDR more vigorously. "Don't you think your candidate should make a burning issue of the confession just made by the President that the country is totally unprepared to make even the slightest defense with land or air forces?" Chamberlain asked Eberstadt in May, 1940. In his next speech, Dewey indeed lashed out at the nation's inadequate air preparedness and chaotic aviation procurement program that left manufacturers utterly bewildered. "This national emergency cries aloud for experts to take complete charge of this program." Dewey responded. "America has the greatest engineering and mass production administrators in the world."[49]

As Dewey campaigned, Roosevelt recruited James Forrestal, that type of outside expert and mass production administrator that the presidential challenger demanded. Baruch intervened on Forrestal's behalf. "Mr. Forrestal has a good reputation and is a quiet but effectual worker," Baruch told the White House. A few days later, Roosevelt brought Forrestal to Washington as one of six presidential assistants provided for under the Executive Reorganization Act of 1939.[50]

Thus began nine tumultuous years of government service for Forrestal. At every stage Eberstadt assisted his friend. By helping Forrestal, Eberstadt served a Democratic administration, at first indirectly and later officially, that he despised. Personal loyalty to his best friend outweighed politics. When Forrestal asked for assistance, Eberstadt gave it willingly and without question. Eberstadt may have had another reason for providing his expert organizational advice to the Roosevelt administration. Perhaps he recognized that through Forrestal he could do his small part to revive the corporate, neo-capitalist structures that had become part of his Progressive and New Era organizational experience but had been buried by the Depression and New Deal. If such corporate organizational impulses did not drive Eberstadt at the moment, the following months of mobilization revealed dramatically that the war crisis created opportunities to graft such organization on an American national security establishment.

3. Mobilization, 1940–1942

Forrestal intrigued Roosevelt. Perhaps, as journalist Eliot Janeway has argued, Forrestal's abandonment of Catholicism or his family's upstate New York Democratic political alliance attracted the president to the Wall Street investment banker. When FDR and Forrestal renewed their acquaintance in early April, 1940, they had reminisced about the old days in 1910 on the hustings in Dutchess County. Forrestal's loyalty to Roosevelt, despite the fact that other Wall Street bankers and conservative Democrats had abandoned the president and were criticizing Forrestal, also impressed FDR. Forrestal had backed New Deal securities, banking, and public utilities regulations but, unlike Eberstadt and most of the Wall Street crowd, had not turned against the social welfare aspects of the New Deal. Corcoran, William O. Douglas, Joe Kennedy, Harry Hopkins, and FDR's budget director, Harold Smith, noted Forrestal's continued fealty to Roosevelt. Forrestal remained useful because he belonged to no particular clique, Smith told FDR.[1]

Forrestal encouraged Roosevelt's early overtures to him. Though he denied seeking a government post and grumbled constantly about the glare of public life, even objecting to his first official photo session in June, 1940, Forrestal's correspondence with Roosevelt in May reflected that of an ambitious man anxious to work with the president. He praised one of Roosevelt's preparedness speeches as a masterpiece and, in another letter, pledged to support FDR's defense program "at whatever tempo you choose to set."[2]

Roosevelt needed a financier who could serve as an intermediary between government and the private banking sector and raise capital for military procurement programs. He wanted an administrative assistant "to help find the men with 'big names' and organizing ability to handle the problem of industrial production of war materials." Forrestal, one administration advisor noted, "has the best list of the new 'big names' in the country." This meant, of course, the Good Man List. FDR also expected Forrestal to mend conservative Democratic political alliances necessary for Roosevelt's reelection to an unprecedented third presidential term. He used Forrestal as he used Henry Stimson and Frank Knox, two leading Republicans brought into the Cabinet as secretaries of war and the navy in June, 1940, to provide an image of national unity for the neutral and divided country. Was it "sheer coincidence," one New Dealer wondered, "or impish Roosevelt delight in dumbfounding the political opposition, to make these appointments right on the eve of the Republican convention?"[3]

Roosevelt employed Forrestal and other recruits from business, industry, and the private sector to maintain executive control over mobilization planning and organization rather than turning authority over to some rival superagency of the type suggested by Baruch. Mobilization brought business, government, and the military together in an intense and uneasy partnership. Roosevelt had to cooperate with private business in order to preserve a tenuous equilibrium between the pluralistic forces and interests that made up American society, about to enter the greatest period of economic and military mobilization in U.S. history. Mobilization meant to FDR the diverting of resources to national defense without interrupting ongoing, peaceful socioeconomic reform begun in 1933 under the New Deal. It required juggling increased war spending and higher taxes with programs for unemployment insurance and old age pensions. Moreover, mobilization caused bitter opposition from isolationists and the fear of big government, cartelization, and dictatorship.

By introducing Forrestal and so-called dollar-a-year men, who, like their World War I predecessors, served without compensation, FDR risked dividing his administration into two divergent political-economic camps with different organizational perspectives. This division underlay the politics of mobilization. Immediately New Deal administrators, bureaucrats, and liberal economists raised the traditional warning of militarism that always accompanied military-industrial cooperation. They worried that the businessmen who arrived in Washington to speed mobilization would ally with the military and seek through corporate structures to wrest control of policy and organization away from New Deal officials. Vice-President Henry Wallace and Leon Henderson, head of the Office of Price Administration, and other staunch New Dealers insisted that the introduction of people like Forrestal into the government united reactionary businessmen, the military, and "economic royalists," who would attempt to remove federal antitrust laws and regulations. Liberals in the Roosevelt administration feared that these mobilizers planned to destroy small business competition and social welfare programs by using the war mobilization process to concentrate economic power as the first stage of multinational cartelization and, through an exclusive managerial elite, to translate their power to government. Liberal journalist Bruce Catton labeled this dollar-a-year elite "the War Lords of Washington."[4]

On the other hand, Forrestal and hundreds of business executives who came to Washington to mobilize the nation saw themselves not as warlords but as organizational experts and professional managers (the type that lay at the base of Eberstadt's Good Man concept). They sought only to provide orderly administration and rational organization. "You did not overestimate the degree of non-organization – not disorganization – and confusion," Eberstadt wrote Clarence Dillon. The professional corporate organizers believed that New Deal bureaucrats had destroyed private initiative and had strangled business efficiency by overregulating and overtaxing. Worse, New Dealers generated conflict among competing economic segments and between labor and management. The dollar-a-year corporate experts thought that New Dealers encouraged wasteful spending and civilian production instead of assisting in the conversion to military production. The frantic New Deal rush toward socialism, the newcomers suggested, forecast a collapse of democratic capitalism and the rise of dictatorship.[5]

While they viewed New Dealers as disruptive politicians, Eberstadt's Good Men and other dollar-a-year types thought of themselves as patriots sent to unite diverse economic interests behind the mobilization program. He came to Washington, Eberstadt informed a group of automobile executives, to prevent the chaos of a "New Deal war" and to protect the "onward march of civilization." He sought to help industrial and business leaders to achieve maximum productive efficiency through industry-wide wage and price stabilization. He hoped to stop labor piracy and teach voluntary pooling of resources. Eberstadt urged the businessmen to control their own interests. "I beg you to put away for the duration any lingering memory that you once were Ford, or Chrysler or General Motors or anyone else," he lectured. "Those tags are obsolete."[6]

Forrestal, too, considered his government service a patriotic, unselfish act, and he divested interest in his Wall Street firm to avoid conflict with government business. Though over time he connected most great Wall Street firms and corporations to government, Forrestal never profited personally from public service. He left Dillon, Read to assist Roosevelt in mobilizing the country, yet once he was on the White House staff of six presidential assistants, Forrestal found little to do. He helped make a motion picture that warned South Americans about Nazi Fifth Column penetration of the Western Hemisphere. He provided liaison with the NBC radio network. Mostly, Forrestal ruminated with naval aviation friends about the growing pressure in Washington for a separate air force department. His job reflected the suspended animation of mid-1940 Washington, which was frozen by a period of neither peace nor war, neither full mobilization nor normal business, and by a president careful not to take action that jeopardized reelection in the fall. The obscurity of Forrestal's position was also due to the built-in anonymity of an office designed by public administration scholars

Louis Brownlow, Luther Gulick, and Charles Merriam along the British model of a cabinet secretary. Brownlow defined the ideal person for Forrestal's job as "somebody who is not a Big Shot, who will not try to become assistant president, but who is skilled in the administrative work of government and who really has a passion for anonymity."[7]

Forrestal chafed at the inactivity. On Wall Street he had been a big shot. As president of Dillon, Read, he had the power and authority to make decisions and select Good Men from Eberstadt's or his own special list. He had neither power nor authority as FDR's assistant. Each day during the summer of 1940 as he disappeared into a White House office, Forrestal heard about the worsening European situation. France collapsed, British forces fled from Dunkirk without their equipment, and German bombers raided Britain in preparation for an expected cross-Channel invasion. At the same time, relations with Japan deteriorated and Axis intrigue increased in the Western Hemisphere. In the face of mounting crises, Congress appropriated unprecedented funds for national defense, including a 70 percent naval increase in July. Roosevelt resurrected the National Defense Advisory Committee of World War I to coordinate this defense program and relied upon the Reconstruction Finance Corporation for stockpiling and conversion to war production. He gave the Army Navy Munitions Board some authority to set priorities for the distribution of scarce materials and finished products. None of these steps seemed adequate to deal with the tremendous military and industrial expansion that accompanied the first phase of mobilization.

Baruch advised the government to employ outside experts to coordinate mobilization. He told Chief of Staff George C. Marshall, an old friend, to hire business executives to organize army contracting and purchasing. At the same time, Secretary of War Stimson recruited organizational experts for his department, including Judge Robert Patterson, Robert Lovett, John McCloy, and others on Eberstadt's Good Man lists. Meanwhile, Frank Knox tried unsuccessfully in the Navy Department to recruit corporate lawyer William Donovan, Chicago oil executive Raleigh Warner, and investment banker Ralph Bard for the new post of undersecretary of the navy. Gradually, Knox's search led to the unhappy, underutilized Forrestal. Apparently Knox did not know Forrestal well, although they shared common oil investment connections and association with Chicago banker and investor Charles Dawes. John Sullivan, Knox's close friend and fellow newspaper executive who in 1947 succeeded Forrestal as secretary of the navy, recommended Forrestal.[8]

Forrestal received the undersecretary's job. No longer buried in an anonymous post, he found himself at the center of the great mobilization. Knox entrusted Forrestal with wide authority over naval material procurement and planning, contracting, tax amortization, and coordination with civilian industry. From his service with the Office of Naval Operations during the first world war, Forrestal recalled the unique naval organization of fleet, bureaus, and shore establishments. He remembered the dominating atmosphere of precedent and tradition in the Navy Department. Familiar faces such as that of his old boss John Towers, now chief of the Bureau of Aeronautics, reminded Forrestal of those earlier days. However, the undersecretary sensed that tradition and precedent had to give way to innovation in the complex new mobilization environment. "I am reluctant to jump at conclusions," he wrote Hitchcock, a former Dillon, Read colleague in August, 1940, "because I realize maybe things have changed since I was in the service."[9]

Forrestal confronted complicated organizational, legal, and administrative problems. He found intense competition for scarce materials, office space, money, and labor. New weapons required research and development programs and wider integration with modern industry. The navy recognized some of these changes. The General Board, founded in 1900 as an advisory body of senior naval officers, told Knox that traditional liaison with the army and industry through joint boards no longer readied the navy for war. "This form of effecting liaison-cooperation is reasonably adequate in peacetime," the board explained in July, 1940, "but is wholly inadequate in national emergency." The navy established a research council in December, 1939, and consolidated the Bureau of Construction and Repair and the Bureau of Engineering into a Bureau of Ships in June, 1940.[10]

For the most part, the naval establishment resisted innovations. Although a fine strategic planner, Chief of Naval Operations (CNO) Harold Stark possessed few management skills. Unlike General Marshall, his counterpart in the army, Stark never fully appreciated the problems of mobilization or business-government relations. At the same time, the chief of the Bureau of Supplies and Accounts insisted that regular naval personnel could handle all the contracting, inventory, inspection, and procurement details without civilian experts. The judge advocate general opposed any increase in the number of outside lawyers to work on contracts, tax questions, and other legal matters, despite the fact that he had no lawyers trained in these fields.[11]

Forrestal turned such limitations into opportunities to construct the undersecretary's office as the chief mobilization agency for the Navy Department. Steadily, he assumed functions, gathered information, and compiled statistics, thus gaining authority over scattered bureaus and offices. He prepared carefully. In late July, several weeks before assuming the undersecretaryship, Forrestal invited Eberstadt and Paul Nitze of Dillon, Read to review his blueprint for industrial mobilization. He planned to leave the naval establishment with its traditions and precedents alone, while broadening his own civilian management base with the recruitment of organizational experts from the Good Man lists.

Forrestal asked Eberstadt for his latest list of "really good names" in plastics, machine tools, aircraft engineering, and various other fields of manufacturing. Eberstadt reminded Forrestal that his Good Man lists remained at his disposal then as ever. Jake Swirbul of Grumman Aircraft, Princeton classmate Frederick B. Rentschler of Pratt & Whitney, Lewis Compton of MIT, and airline entrepreneur Nicholas Ludington, among others, could get things started on aviation procurement. Badly needed legal specialists Charles F. Detmar, Jr., and William Dulles were available from Eberstadt's old law firm and from Dillon, Read's current counsel, the firm of Wright, Gordon, Zachry and Parlin. H. Struve Hensel, a Princeton and Columbia law graduate associated with Milbank, Tweed and Hope, possessed expertise on tax amortization and contracting. Clarence Dillon's son Douglas, vice-president of the firm and a naval reservist, and Ferol Overfelt, a retired naval officer now with National City Bank, could study the organization of a statistical service. Lewis Strauss of Kuhn, Loeb might review the navy inspection system. Merchandising executive Frank M. Folsom, one of Eberstadt's best friends, provided expertise on industrial liaison. For his part, Eberstadt agreed to survey the entire machine tool industry.[12]

With the assistance of these and others from the Good Man lists, Forrestal placed his people at the interstices of power within the naval bureaucracy. He attached legal experts, statisticians, and inspectors directly to bureaus and offices. Centralization through the undersecretary and executive management structures such as offices for statistics, inspection, procurement, and material provided overall coordination and policy control as in a modern corporation. Agents of the undersecretary, who scattered about the decentralized naval bureaucracy, became the interconnecting threads of a corporate system. Samuel Robinson of the Bureau of Ships and management-oriented officers in the Bureau of Aeronautics actively supported Forrestal. Naturally, the old organization resisted loss of authority. The judge advocate general lobbied to cut appropriations for more than one legal expert and the Bureau of Supplies and Accounts placed numerous subtle obstacles in the way of Forrestal's operations. In the end with the aid of sympathetic naval officers, outside experts, and corporate organizers, Forrestal gained control over the navy's industrial mobilization process. "The situation had become bogged down a bit but now, thanks to your help," he wrote Eberstadt, "is back on the track apparently finally and for good."[13]

Forrestal's ability to organize the industrial mobilization program and to centralize executive authority and policy in the undersecretary's office impressed FDR's budget director, Harold Smith, a Brownlow student of modern government administration. Smith found Forrestal's management techniques superior to those of nearly every other administrator in Washington. Unlike the executives in the Office of Production Management (OPM) and other emergency agencies built one on top of the other throughout late 1940 and 1941, Forrestal showed the managerial skills to facilitate government-industry relations. Forrestal was a "cold, curt, incisive type," one OPM official marveled. He persuaded factory owners through tax incentives, advance ordering, machine tool pooling, and labor-management cooperation to convert to war production for the navy. By contrast, the War Department could not get Chrysler Motors president Kaufman T. Keller to put production of military vehicles ahead of the rapidly recovering civilian automobile line. The frustrated chief of staff, General Marshall, wrote Keller that the army could not complete its maneuvers without such contracted equipment.[14]

Smith promoted Forrestal for the position of "director of economic defense," and his recommendation carried considerable influence with the president. Smith, the seemingly nondescript former budget director for the state of Michigan, had become a major force in the administration. Roosevelt called him "Mr. Fix-it." Brought into the government in early 1939, Smith had recruited top public administration experts and molded his bureau into a central executive management and coordinating agency for the president. Smith shared Roosevelt's views of an activist federal government and strong managerial presidency. He sought to protect and strengthen both through his own mediation and by facilitating communication between the White House and Congress, the press, and executive departments. Mobilization threatened to undermine presidential authority as pressure grew for a central mobilization agency and a mobilization czar who would be a second president. Smith wanted to block this movement by developing his own office of economic defense with Forrestal as the director.[15]

Smith told the president that Forrestal had the drive, courage, and independence from political interests to become their director of economic defense. He urged Roosevelt to appoint such an executive because the piecemeal mobilization program eroded presidential authority. Smith thought that the president's formula of balancing interests through partial measures and temporary agencies no longer sufficed. The recent Lend-Lease law further complicated mobilization by adding foreign consumers to American civilian and military customers. Roosevelt's approach defied theories of executive control and centralization espoused by Smith's consultants and budget experts, such as Arnold Miles and Leonard W. Hoelscher, who had learned concepts of administration from political scientist Leonard White's "neo-Hamiltonian" school, or from Brownlow, Gulick, and John Millett at the Public Administration Clearing House (PACH) in Chicago. More important, Smith worried that if the president did not create an executive director soon, he would lose authority over mobilization, which both men insisted must remain with the White House.[16]

Already, Warren Austin, a member of the Senate Military Affairs Committee (and the only member of Congress on Eberstadt's Good Man List), had complained about the president's failure to develop central responsibility and decentralized operations for industrial mobilization similar to that of efficient modern corporations. In March, 1941, the Senate organized a committee under Sen. Harry Truman to investigate production bottlenecks and other executive mobilization problems. Reminiscent of a similar committee that had probed Wilson's war mobilization program in 1918, the Truman Committee clearly challenged Roosevelt's executive powers, particularly his wholesale employment of dollar-a-year assistants. Critics inside and outside the government increased attacks on the president. General Marshall told Senator Truman that the president's mobilization program resembled "an Indian-war plan." Baruch criticized the president publicly for disorganized pricing, priorities, and production policies. The observations of Edward Mead Earle, a defense expert with the Institute for Advanced Study, were typical. "The truth of the matter is, as I see it, that the President is the biggest single bottle-neck in matters of production, strategy, and morale," he wrote journalist George Fielding Eliot. "He is, I fear, a tired man whose inertia must be overcome."[17]

Smith advised the president that the answer to such criticisms lay in appointment of a director of economic defense. Hopkins and Dean Acheson, undersecretary of state for economic affairs and architect of the legal aspects of the Destroyers-for-Bases deal in September, 1940, agreed that Forrestal could do the job. They considered Acheson for the position, but "this brought a response of modesty from Acheson who said that he was not only the world's worst administrator but was no good in pushing people around," Smith told the president, adding that Acheson "said that he thought Forrestal was the best person for the position of Director of Economic Defense."[18]

Roosevelt ignored the recommendation, preferring to leave Forrestal in the Navy Department to carry out administration preparedness, much as FDR had done as Josephus Daniels's assistant during the first world war. Eliot Janeway thought the president kept Forrestal in the Navy Department as his inside source of information and control. "It was the one place, in fact, in which Roosevelt felt obliged to put a man unusual enough to operate as he himself had operated as a young man," Janeway observed. More likely, Roosevelt did not want to create administrative competition or bring to a head through such an appointment the increasingly intense struggles between those who wanted complete mobilization and those who preferred business consolidation and recovery from depression with a minimum of government coercion and regulation.[19]

Establishment of "an all powerful head similar to the so-called czars of baseball and the moving picture industry," particularly a czar such as Wall Street speculator Forrestal, promised to sharpen the emerging conflict between New Deal officials and the dollar-a-year corporate experts. The concept of an economic czar was, one New Dealer warned, a step on the "road to full, permanent cartelization of industry." The czar would embody the values of conservative business leaders who wanted control by a managerial elite and board of corporate directors. It suggested as well a corporatist solution similar to that proposed by some reactionary (even fascist) foreign regimes. Indeed, Eberstadt raised this image in early 1941. Eberstadt told Forrestal that a political-economic czar and corporate board of mobilization directors reminded him of Metternich and his Congress of Vienna that rebuilt Europe after the destructive Napoleonic Wars. A contemporary congress of experts could plan for the current "problem of turning over into productive use and occupation at the end of the war, the hundreds of millions of people and the mass of facilities which has been employed in nonproductive use." Eberstadt continued: "It is a job similar to what was done at the Conference of Vienna [1815], and while it has become fashionable to hold the results of those old fellows in contempt, the fact is that they did produce peace practically throughout the whole world for a period of about a century."[20]

Whether framed in a traditional conservative or quasi-corporatist setting, the idea of a great mobilization czar and corporate board repelled Roosevelt. Nevertheless, he could not ignore mounting pressure for the appointment of some mobilization coordinator and the construction of a central agency for economic defense. FDR decided in August, 1941, to name Donald Nelson the executive director of a new Supply Priorities and Allocations Board (SPAB). Nelson was an executive with Sears, Roebuck and an NRA veteran. "He was big, jovial, and self-possessed, and his round red face beamed through the cigar smoke," one aide remembered fondly. Friendly and more relaxed than the ambitious corporate chiefs and dollar-a-year management types that dominated the mobilization machinery in 1941, Nelson appeared to be the ideal moderate businessman, with experience heading "the biggest buying organization in the United States," to run a central mobilization agency. He had none of the czar complex apparent in others. "There was always a faint New Dealish flavor to Nelson," an assistant recalled. Nelson sought the advice of liberal economists such as Robert R. Nathan and became intimate with staunch New Dealer Leon Henderson. Most important, Nelson "was known to think FDR a great man."[21]

Nelson and SPAB received only vague policy-making guidelines. SPAB held no clear authority over other agencies. It lacked interorganizational links or coordinating machinery. Eberstadt, for one, called it nothing more than a "hydra-headed monster." It was a transitional agency, though, and a stage in the creation of what the president termed the great "arsenal of democracy." "In spite of its limitations," Nelson insisted, "SPAB was instrumental in paving the way for the tremendous production program that was to follow."[22]

While SPAB struggled to pave the way for the arsenal of democracy, individual mobilization organizations developed their own machinery. Much of it lay closer to the corporate thinking of dollar-a-year experts than to that held by the New Dealers. Nowhere was this corporatization process more evident than in the Navy Department, where Forrestal, supported enthusiastically by Secretary Frank Knox, forged ahead with his own innovations for civilian control of the naval bureaucracy. These innovations included offices and divisions for administration, scientific research and development, contract distribution, and a procurement legal counsel, a position filled by Wall Street attorney H. Struve Hensel. In September, Artemus ("Di") Gates, a New York banker, insurance executive, and director of Pan American Airways, became assistant secretary of the navy for air, a post vacant since 1933. Gates, who was once Forrestal's roommate in their bachelor apartment, joined Hensel on the inside and Eberstadt on the outside to form what was essentially the undersecretary's corporate board of directors. This alliance enabled Forrestal to become a formidable factor in the bureaucratic battles over mobilization, wartime defense production, and postwar national security organization.

Forrestal needed more than structural innovations, however. Without a central coordinating agency and interorganizational mechanism, he found it difficult to define and balance navy demands for production with those of the army, army air force, Lend-Lease borrowers, and civilian market. He discussed these problems throughout the spring of 1941 with Eberstadt in correspondence and over the telephone. The latter proved to be an unsafe means of communication. "Somebody must have been listening in," he cabled Forrestal after one telephone conversation. Partly to avoid leaks, Eberstadt traveled to Washington regularly between May and July to consult with and advise Forrestal. "I may have to go down there full time sooner or later," Eberstadt wrote William Chamberlain.[23]

Though he went to Washington with the limited goal of assisting in emergency organization, the mobilization process soon stimulated Eberstadt's corporatist inclinations, which he discussed with Forrestal. Forrestal shared his friend's assumptions but apparently never developed a clear corporatist philosophy of his own. Nevertheless, both recognized early that partial mobilization provided an opportunity to reorganize American production. Eberstadt expected mobilization machinery to unite labor and capital, government and business, and taxpayer and bureaucrat in a national effort. He insisted that with firm determination by the administration and a properly organized program, the United States could produce 500,000 to a million warplanes a year. Such production, Eberstadt told Forrestal, would have a "very sobering effect on the bandits" and restore American confidence shattered by the Depression.[24]

How should the national economy and production be organized? No one had a political-economic blueprint. "The present world situation is due to the fact that neither Democracy, nor Fascism, nor Bolshevism, nor Nazism for that matter, have been able to use their equipment in such a way as to sustain and raise the standard of living of their respective populations," Eberstadt told Forrestal. "The hard fact boys" of the New Deal had not found the answer. They lacked creative power to organize production and provide national unity. "I feel confident that this war cannot be won through conventional thinking and reasonable efforts," Eberstadt stated. He offered Forrestal no political-economic plan of his own at the moment but pledged to do his part "when the time and opportunity presented themselves."[25]

That opportunity appeared sooner than expected. In June, Forrestal and Undersecretary of War Robert Patterson replaced the assistant secretaries as the civilian representatives on the Army Navy Munitions Board. Both thought that despite the board's marginal contributions in the past, this neglected joint agency contained the machinery for interorganizational coordination. Founded by executive order in 1922 as an outgrowth of the mobilization experience during the first world war, the ANMB was placed in the office of the assistant secretary of war according to the provisions of the National Defense Act of 1920. There, it collaborated with the Army Industrial College, founded in 1924, to study the problems of mobilizing civilian production for war, and it helped develop a series of industrial mobilization plans. Baruch, Hancock, Hugh Johnson, and other War Industries Board veterans, civilian business and industrial leaders, and experts from the Harvard Graduate School of Business Administration met with army planners (including Dwight D. Eisenhower in 1931) to develop mobilization plans. Consultation remained casual and sporadic, and the mobilization plan that developed reflected concern for civilian control over the mobilization process. Constrained by popular revulsion to munitions makers and militarism such as that embodied in the Nye Senate investigation of the munitions industry in 1934, civilian and military planners developed a moderate program centered on employment of the ANMB as a temporary central agency to ease the transition from peace to war until civilian authority could assume direction of mobilization.[26]

The industrial mobilization plans showed few characteristics of a nascent military-industrial complex. Navy collaboration with army and industry planners remained minimal. Unlike his War Department counterpart, the assistant secretary of the navy held no statutory membership on the ANMB. Moreover, the navy's industrial mobilization requirements and own internal procurement organization differed so much from that of the army that it saw little need to develop joint industrial mobilization plans. The naval planners on the General Board and in the bureaus and offices of the naval establishment focused instead on fleet operations, maintenance, and strategic planning. Navy participation on the ANMB increased when the board reorganized with an executive committee in 1931, and naval officers began attending lectures at the Army Industrial College. Even then, the Navy Bureau of Supplies and Accounts–Forrestal's nemesis in 1940–41–refused to participate in revision of the industrial mobilization plan.[27]

During the late 1930s, Assistant Secretary of War Johnson used the ANMB and the industrial mobilization plan in his struggle with Secretary of War Harry Woodring, an isolationist, to move the administration toward preparedness. Johnson pressed war planning on the president. In response, Roosevelt requested an updated industrial mobilization plan and convened a War Resources Board to facilitate planning. Continued concern that such measures represented a "fascist blueprint for war" forced Roosevelt to ignore the War Resources Board. Moreover, he refused to employ the ANMB as a central coordinating agency. Roosevelt appeased opponents of intervention but frustrated Baruch and other advocates of industrial preparedness.[28]

Finally, in July, 1939, Roosevelt placed the ANMB directly under presidential authority. During the next two years the board struggled to define its own authority over priority ratings, relations with the National Defense Advisory Committee, Lend-Lease, and other organizations. Few consulted its vast storehouse of information on factories, transportation, power supplies, and strategic materials, until Forrestal and Patterson (probably advised by Baruch) decided in August, 1941, to employ the Army Navy Munitions Board as the central agency for interservice coordination. They used the board to link the military with industrial, business, and financial sectors of the economy. The undersecretaries wanted the board to define and balance civilian production with the demands of the army and navy. They recognized that the ANMB needed reorganization and direction by a civilian manager rather than by the assistant secretary of war, as provided by statute.

Forrestal and Patterson asked Eberstadt to study the ANMB and recommend reorganization. Thus, world crisis and the accelerated pace of mobilization gave Eberstadt an opportunity to develop political-economic strategies and innovative structures. Though not apparent at the moment, the ANMB study, which went well beyond his informal machine tool study he had prepared for Forrestal earlier in the year, became the first phase of a series of reports, studies, and organizational innovations that between 1941 and 1949 Eberstadt contributed in the process of defining the first modern American national security system. Eberstadt's correspondence with Forrestal in mid-1941 indicated that he thought about wider political-economic issues when he discussed the mobilization process and that the ANMB job provided a way to translate that thinking to government organization.[29]

The ANMB project marked Eberstadt's transition from private businessman to public servant. He asked Forrestal whether he should divest his holdings in chemical companies and other businesses with national defense contracts. Like his best friend, Eberstadt never profited directly from the mobilization crisis but made certain that the right people from the biggest banks, insurance companies, and corporations contacted the appropriate government agencies. Starting his ANMB survey in October, 1941, Eberstadt discovered familiar names–many already on his own Good Man List–making the adjustment from private business, law, and industry to the dollar-a-year posts. Those who figured most prominently on Eberstadt's list included John McCloy, Artemus Gates, H. Struve Hensel, Robert Lovett, John Lord O'Brian, William Donovan, Floyd Odlum, and Donald Davis. Each operated as a legal, management, or production consultant with various governmental departments and agencies.

Predictably, Eberstadt's first interviews for his ANMB study came from those who fit the Good Man image. He visited corporate lawyer Stuart Nash Scott, a naval reserve officer on the ANMB. Then, he talked with accounting executive Arthur H. Carter, a member of Patterson's staff at the War Department. Eberstadt also consulted Stacy May, a Rockefeller Foundation statistician who, as an ANMB expert, was actively involved with longtime Eberstadt business associate Jean Monnet in defining foreign production requirements for the OPM and the French and British purchasing commissions.[30]

Stuart Scott typified the organizational expert found on Eberstadt's Good Man lists. A Harvard law graduate, Scott belonged to the firm headed by Tom Dewey and Arthur Ballantine, who were Eberstadt's friends and Wall Street neighbors. An expert on corporate and financial law, Scott had earned a reputation for "apolitical" dedication to reconciling differences between interests as counsel for the French (and later British) Purchasing Commission under Monnet. Scott also worked for the Navy Bureau of Aeronautics, the ANMB, and later as general counsel for the Surplus War Property agency.[31]

In many ways carbon copies of Eberstadt at Cotton and Franklin in the early 1920s, Scott and Good Man lawyers such as Hensel, Charles Detmar, and William Dulles in Forrestal's office and McCloy, Edward Greenbaum, and Goldthwaite H. Dorr with the War Department held the same values and spoke the same language of corporate adjustment, institutional reorganization, and intermediation between interests. During Eberstadt's years of wartime and postwar national security study and organization building, such lawyers provided constant support. Together, they advanced interservice, interagency, civil-military, and government-business cooperation. War Department counsel Dorr, who carried on his own inside studies, praised Eberstadt's efforts to rationalize the ANMB. "It is certainly a record of foresight and activity," Dorr told Patterson. "It is unfortunate that many of these recommendations did not receive earlier acceptance."[32]

Eberstadt found familiar views as well in conversations with statistician-economist Stacy May. A product of the Rockefeller-financed Brookings Institution, May held many of the same political-economic theories that Eberstadt had developed with Fetter at Princeton, McAdoo and Cotton on Wall Street, and Young on the reparations commission. Indeed, one of the Brookings Institution's founders, W. F. Willoughby, had joined Fetter's department shortly before Eberstadt's graduation from Princeton in 1913. Together they viewed nonpartisan research such as that provided by the Brookings Institution, or through studies by outside experts such as Eberstadt, as an important tool in asserting bureaucratic control over individual interests. They sought to separate administration from politics through the employment of experts in business, finance, industry, accounting, education, and law. This managerial elite would develop with government a rational, efficient, and balanced economy of plenty. Brookings experts wanted state regulation kept to a minimum. The state, through such experts and the development of intergovernmental agencies, acted as honest broker. It coordinated the many conflicting interests. Ideally, it stabilized through balanced budgets, stimulated international commerce, and eliminated tariffs. It controlled prices and production.[33]

The war mobilization emergency provided economists like May with an opportunity to apply some of these principles to what he told Eberstadt would be a "Victory Plan." May showed Eberstadt a chart that outlined the requirements of the many governmental and private consumers against the available resources. Balancing wartime demand with supply became their most urgent concern. They had to build mechanisms that brought requirements into line with production. Already the army's requirements, May told Eberstadt, had exceeded available supplies by 100 percent. May warned that "this clearly indicated a definite and violent collision of the various contenders for the limited amount that would be available."[34]

In his conversations with May, Eberstadt considered the problem of how to control the flow and distribution of scarce critical materials to various claimant agencies. He studied how to balance production with policy during war mobilization and May's charts as a blueprint for postwar economic and national security organization. Conversations that same day with former ANMB secretary Harry K. Rutherford, a veteran army ordnance officer with a master's degree in business administration from Harvard, crystallized Eberstadt's thinking. Rutherford had considered the problem of mobilization for years. As dean of the faculty of the Army Industrial College, he had rewritten the curriculum to reflect the importance of a military understanding of the civilian economy and industrial organization. He had invited Baruch and other WIB veterans to share their mobilization wisdom. Between 1933 and 1938 Rutherford developed an industrial mobilization plan for the New York Ordnance District and advised Assistant Secretary of War Louis Johnson, the ANMB, and the War Resources Board. Rutherford, a Republican and an expert on investment cycles, influenced Eberstadt's entire approach to mobilization planning. He told Eberstadt that answers to the current organizational and mobilization dilemmas could be found in a history of the War Industries Board entitled *Industrial America in the World War*, by Grosvenor B. Clarkson. Rutherford located eight specific page references in this book where Eberstadt could find material to guide his study.[35]

Clarkson had prepared his book in 1923 as a promotional tract to keep alive in peacetime the mobilization lessons of the first world war. He had publicized the organizational philosophies of WIB director Baruch, who had paid for the study and its publication. The rhetoric of economy, efficiency, voluntarism, and self-regulation from the liberal, democratic capitalist images of the Progressive Era filled Clarkson's work. He presented the War Industries Board as the ideal structure, a perfect, decentralized administrative machine run by industrial, business, and academic experts. No more than fifteen hundred of these managers organized "order, reason, and restraint" into the entire war production process "and imbued the huge economic mechanism of America with potent efficiency." By employing a system of priorities and price control, Clarkson wrote, the WIB met all military, naval, and civilian demands with as little dislocation of industry as possible.[36]

Could fifteen hundred management experts confronted by the complex organizational demands of 1941, and by material scarcity rather than the abundance of the first world war, manage mobilization through a voluntaristic price and priority system? Roosevelt's administrative expert, Luther Gulick, maintained that such methods were obsolete and could not solve current problems. "We learned little about industrial mobilization and war economic institutions from World War I," Gulick insisted. Eberstadt disagreed. Clarkson's study reflected Eberstadt's own Progressive Era educational values. Moreover, it instructed him in the fundamental problems and methods of industrial mobilization and economic institutions at war. Eberstadt borrowed specific ideas from those page references noted by Rutherford and translated these to his ANMB report. He later applied them to the War Production Board, Controlled Materials Plan, Unification Report, and national security studies and legislation. There would be far more institutional continuity than Gulick imagined.[37]

In theory, the WIB provided a corporate structure to organize the American economy in war and peace. It served as a policy-centralizing clearinghouse and coordinating agency linking government to industry. Industrial divisions and commodity sections, each run by a director, board of experts, and government representative from the purchasing agency interested in that particular commodity, exerted control over conflicting interests. Division directors authorized priorities and contracts. They translated central policies to private industries through regional advisors and the war service committees of the Chamber of Commerce. Thus, the WIB stabilized production with a minimum of federal coercion, Clarkson suggested. "The few thousands of industries located at the source of the *controlled materials* were so grouped together by association or committees, Clarkson wrote, "that they effectually policed themselves and their operations were so conspicuous that any glaring evasion or violation of regulations for any length of time was impossible [emphasis added]."[38]

Eberstadt knew that such a voluntary, self-regulating association, if it had ever existed in this form at all, did not apply to the current mobilization process. However, he accepted the WIB's corporatist concept of industrial committees and commodity sections licensed by and coordinated through the state to control materials. He adopted WIB principles that the contracting process balanced supply and demand, influenced production quotas, and eliminated conflicts among industries and between labor and management. WIB substructures–run by expert businessmen–"represented the highest type of administrative organization the United States has ever seen," Clarkson argued. "The commodity sections were business operating government business for the common good." By turning on and off the flow of critical materials from a central administrative "faucet," the WIB balanced supply and demand, stabilized production, standardized methods, and eliminated damaging competition. Thus it protected the fundamental democratic-capitalist system. "Russia was torn to pieces by its Soviet proletarians," Clarkson concluded in 1923, and "America was united by the protean forces of its managers and producers."[39]

Eberstadt absorbed Baruch's philosophies and methods in reading of Clarkson's tract. A visit with the mobilization sage himself seemed mandatory. Eberstadt renewed acquaintances. He had met Baruch during the Versailles conference at the end of World War I, and since the early 1920s he had known the Wall Street speculator well through McAdoo, Dillon, Swope, and J. Leonard Replogle, the latter Baruch's WIB Steel Division director and Eberstadt's legal client. They held a long and friendly discussion bonded by their mutual admiration and respect for Forrestal, Patterson, and William Douglas, who they agreed should have received presidential appointment as sole coordinator of mobilization.[40]

This would be the first of many Eberstadt-Baruch consultations during and after the war and the start of a friendship that lasted for the remainder of Baruch's long life. Together they battled New Deal bureaucrats, served on the UN Atomic Energy Commission, and put their organizational views into practice with the aid of dozens of mutual friends and associates. In many ways the energetic Eberstadt seemed to be a younger version of Baruch. Like Baruch, Eberstadt moved between the layers of power, mediating relations between industry and military, business and government, academic experts and government bureaucrats. Baruch handed a corporatist legacy to Eberstadt, and the latter never disappointed him.

Eberstadt needed to learn the inner workings of the bureaucracy before he could link its parts or tie it to the private sector. He found little business management wisdom in the regular naval establishment in 1941. He considered CNO Stark pleasant but ill-informed on economic organization. At the same time, one bureau chief spent his time with Eberstadt explaining the battle against the *Graf Spee* rather than his procurement responsibilities on the Army Navy Munitions Board. "His information on it [the board] seemed just a little bit less than his interest in it," Eberstadt observed. Even naval officers committed to the smooth functioning of the ANMB committees showed organizational ignorance. "Captain Anderson thought that Commander Henning was under Captain Almy," Eberstadt observed, "but I have the impression that Henning is in the Under Secretary's Office."[41]

Relations between army and navy representatives on the board appeared good. They reflected the cooperation between Forrestal and Patterson and between army and navy reservists with similar legal, financial, and business backgrounds and values. Snags in production coordination stemmed not from interservice rivalries or feuds over scarce supplies but from lack of policy definition and clear organizational lines. The board failed to link army and navy to the larger mobilization plan because of Roosevelt's approach, Scott told Eberstadt. "Commander Scott showed me a letter from the President, turning down a request to the Joint Boad on the matter of general authority to set mandatory priorities on matters concerning the Army and Navy." Surprisingly, Roosevelt partisan Donald Nelson, executive director of the Supply Priorities and Allocations Board, agreed that the Army Navy Munitions Board had not been employed properly. It should become an "all-out advocate of the military services," Nelson told Eberstadt, so long as it was placed under a civilian director. Already forecasting the bitter rift and bureaucratic struggles that accentuated the Eberstadt-Nelson relationship during the war, Eberstadt found it amazing to hear him express views that basically paralleled his own.[42]

Eberstadt insisted on civilian leadership for the ANMB and control over the military. He emphasized civilian control in his final report to Undersecretaries Patterson and Forrestal. Such control resided in the chairman and four-man executive committee composed of the undersecretaries of war and navy, an army supply officer, and a navy procurement officer. His report urged the closest coordination between ANMB staff and civilian mobilization agencies, particularly the Office of Production Management. Eberstadt visualized the board's function as the determination of exact military and naval needs and the translation of those requirements to industry.

Instead of creating an agency for the military to exert control over the economy, as liberal New Deal critics later averred, Eberstadt employed the board to ensure that army or navy demands for materials and production did not exceed supply. The problem of excessive demand had been revealed to him by Stacy May. Eberstadt envisaged the ANMB as a mechanism through which Patterson, Forrestal, and civilian administrators divided limited resources between the services, and provided civilian-military cooperation and industry-government coordination. In the end, he returned to the basic industrial mobilization plan that made the ANMB the transitional coordinating agency until civilian structures assumed responsibility for mobilization. Privately, he expected board reorganization to pave the way for a great civilian War Industries Board, modeled after the ideal coordinating agency presented in Clarkson's 1923 promotional tract.[43]

Patterson and Forrestal welcomed the ANMB report, which coincided with their own views. Eberstadt, in effect, had supplied them with an administrative tool to help them persuade their bosses, Stimson and Knox, to press the president on reorganization. Such structural adjustment became more urgent a few days after Eberstadt submitted his final report; it was then that the Japanese attack on Pearl Harbor catapulted the United States into the war. On December 22, 1941, Secretaries Stimson, Knox, and their undersecretaries asked the president to accept the Eberstadt proposal. That same week, Roosevelt's top mobilization executives reviewed the proposal and urged the president not to adopt it. They saw the report as a thinly disguised version of Baruch's single mobilization czar scheme that Roosevelt had resisted since late 1938. OPM Director William Knudsen opposed the reconstituted ANMB because it duplicated his own agency. Nelson moderated earlier enthusiasm expressed to Eberstadt for an "all-out" organization and now voiced reservations concluding that the report created an organization for military control over industry. Leon Henderson, at the Office of Price Administration, was more direct. He claimed that Eberstadt sought to place the entire industrial life of the nation under military domination.[44]

Eberstadt's ANMB report sharpened the lines of division between that inner circle of dollar-a-year Good Men and the New Deal public administrators. Roosevelt could ignore neither the entreaties of his service secretaries to adopt Eberstadt's plan nor the opposition to the proposal by his own New Deal officers. Above all, the president needed to avoid a decision that forced confrontation between the major groups essential to his political and economic conduct of war mobilization. As had become his organizational approach during the years of neutrality, FDR pacified both groups, pleased neither, and delicately balanced forces (and enhanced presidential power) by creating more competing structures. He placed Vice-President Wallace in charge of a Board of Economic Warfare (BEW) to deal with foreign governments on export-import policies and procurement of strategic materials. Then, Roosevelt selected Nelson, seen by some as the liberal New Deal friend of small business and labor, to head the War Production Board (WPB), which would have authority over industrial organization for war, supply priorities and allocations, and domestic mobilization policies. Theoretically, Nelson became the new Baruch. Simultaneously, Roosevelt approved the reorganization of the ANMB and, on the advice of the civilian leaders in the War and Navy departments, made Eberstadt the ANMB's first civilian director.[45]

With Eberstadt's appointment to the eight-thousand-dollar-a-year position, Roosevelt tilted his delicate bureaucratic balance of power toward the corporate organizational managers. In the months ahead, Eberstadt—assisted at every step by Forrestal, Patterson, and Baruch—drove ahead relentlessly to graft his corporate organizational theories and structures onto government. Eberstadt had not sought this position, but he did enter the job fully aware that mobilization advanced public-private partnership. Presumably as a patriotic, nonpartisan volunteer, he did not expect his presence in Washington or his innovative organizational ideas to create conflicts with New Dealers or other members of the regular administrative machinery of government. Nevertheless, conflict and controversy plagued Eberstadt's mobilization job. In this struggle, Nelson became Eberstadt's foil. For a time the WPB director resisted, then hired, and ultimately fired Eberstadt. In the end, the outcome of this classic bureaucratic fight over mobilization and political-economic philosophy lay in the fate of Eberstadt's Controlled Materials Plan.

4. Controlled Materials, 1942–1944

The dollar-a-year mobilizers seemed to thrive on the war crisis. "Everybody is telling us now we must be tough," said David Lilienthal, chairman of the Tennessee Valley Authority, observing that "the other day an Assistant Secretary of the Navy, whose calluses have all come from polo, put on a talk that sounded as if he were an ex-stevedore." Eberstadt added to the tough talk. "We meet tonight in one of the grimmest moments of the darkest hour of modern times," he told an assemblage of Michigan industrialists on July 3, 1942. Hong Kong, Singapore, Burma, the Malay Peninsula, the Philippines, and the Dutch Indies all lay in enemy hands. Poland, Belgium, Holland, Norway, France, and "a tenth of Russia" had fallen to Hitler's "blitz-buggies," he continued. "We've lost valuable Mediterranean bases. We've lost islands of strategic importance off Alaska. We've lost the battle of the tanker lanes on the Atlantic seaboard and in the Caribbean. We've lost our rubber supply and most of our tin," he stated. The only thing that America had not lost, Eberstadt concluded, was its complacency.[1]

Eberstadt attacked that complacency. He exhorted that group of machine tool and automobile executives to convert more rapidly to full war production. He urged them to cooperate with labor and to share resources. Above all, Eberstadt asked them to put aside their individual interests in the face of mounting national crisis and thereby create a "free world" and assist the "onward march of civilization." "We're all in the same boat," Eberstadt preached, "management and labor, Democrats and Republicans, easterners and westerners, rich and poor, black and white, Protestants, Catholics, and Jews." He closed the speech with a warning: "What we are talking about is life and death itself, life or death for individuals, life or death for our institutions, our country."[2]

That same intensity, that same quasi-religious crusade for unity, production, and national harmony, drove Eberstadt at his ANMB post. He assaulted routine, drove through programs, and built new structures. "I went right straight for my objective and probably stepped on a lot of toes which maybe I didn't need to step on, but when they got in the way, they got pushed aside," he admitted. Fellow ANMB executives Forrestal and Patterson displayed similar single-minded pursuit of full mobilization. They struggled to curb antitrust measures that interfered with production and resisted regulations that hindered private industry. The ANMB team advanced credit and tax policies that would assist business expansion. The three civilian mobilizers joined Army Service Force (ASF) generals Brehon Somervell and Lucius Clay, Adm. Samuel Robinson, and other top military and naval procurement officers to obtain maximum war production. They became such a close and relentless team that presidential assistant Eugene B. Casey claimed they formed a cabal that sought "a fascist military dictatorship for the United States."[3]

Though driven to provide the military everything it required to fight a world war, the ANMB never comprised a military cabal. It moderated excessive production demands by its military clients, balanced army and navy interests with the overall war effort, and in the words of Schwarz, "prevented the Army from running amok." Forrestal battled Adm. Ernest King's attempts to centralize more authority in the office of the Commander-in-Chief. Simultaneously, the undersecretary bypassed naval supply bureaucracy while constructing his own civilian machine through such innovations as the Procurement Legal Counsel and the Office of Procurement and Material. Patterson had more difficulty in defending civilian control since the Marshall reorganization of 1942 placed many of his functions with the Army Service Force under General Somervell. Nevertheless, he remained an independent administrator, too. "Somervell's principal objection to Patterson," noted ASF counsel Edward S. Greenbaum, "was Patterson's complete disregard of military channels."[4]

The civilian management team on the ANMB shared a common perspective with military officers on the problems of mobilization, though. Both the staff and the officers found the New Deal lawyers, statisticians, and economists on the War Production Board more interested in reforms than in mobilization. Both thought WPB director Nelson weak and incompetent. Both agreed that while the WPB measured the feasibility of military production goals, it delayed the production process needed to meet those goals. Civilian and military mobilizers worried that WPB administrators relied upon the Production Requirements Plan (PRP), based on a World War I priority rating system that caused consumer agencies to exaggerate their demands in the scramble for scarce raw materials. Consumers hoarded supplies and created production bottlenecks. Priority ratings did not address the need for correlation between raw materials, component parts, and finished products. The PRP had no central control mechanism for the mobilization program and no direct links between purchaser and the manufacturer. Contractors who sought material priority ratings flooded the WPB with fourteen thousand applications a week. "Well, for God [*sic*] sake, such a paper avalanche as you'd never believe," Eberstadt cried. It had become, in his words, "industrial guerrilla warfare" around the nation and, back in Washington, name calling and bitter exchanges between Somervell and Nelson's WPB liberals.[5]

Eberstadt tried to mediate between these interests and factions. He hoped to secure cooperation between groups, just as he had done as corporate counsel for Cotton and Franklin and for Dillon, Read in Europe. This time, though, he lacked either the instrument or platform for such adjustment. The ANMB oversaw only military procurement and interservice coordination. Its authority was strictly defined and in many ways was too limited. It became neither the vehicle for overall economic coordination nor for the promotion of larger politico-economic reorganization. Moreover, during peacetime periods of economic abundance, money provided interorganizational facilitators like Eberstadt with a determining mechanism for institutional arbitration by attracting material, labor, and business. Money balanced diverse interests in peacetime; in wartime the controlling factor became raw materials.

Eberstadt thought seriously during his ANMB study in late 1941 about the Controlled Materials Plan (CMP) concept of a central coordinating mechanism. Rutherford had directed his attention to a specific controlled materials idea in Clarkson's WIB history. Each day on the ANMB Eberstadt observed how every bottleneck in production, problem in supply, and conflict of interests could be traced to the inability of the priority system to link the three materials that were the "heart of production"–steel, copper, and aluminum–to specific manufacturers. America's most successful corporations had overcome such problems by instituting a vertical material allocation plan. Alfred P. Sloan, Jr., at General Motors solved production bottlenecks, Eberstadt observed, by tying available materials directly to production forecasts all the way up the line to final output. GM thus became in Eberstadt's view "the best organized company in the United States."[6]

In March, 1942, Eberstadt suggested to William L. Batt, WPB vice-chairman, that the board consider his Controlled Materials Plan idea. Batt ignored the suggestion. A few weeks later, Eberstadt sent the proposal to James S. Knowlson, head of the Division of Industry Operations. Apparently, Knowlson did not understand the concept but did forward it to Robert Nathan, WPB planning chief and noted expert on national income. The liberal economist found enough merit in the proposal to urge Nelson to hire Eberstadt as Knowlson's assistant. Nelson actually made such a job offer in June. Recognizing that this position lacked authority or influence, Eberstadt declined. He held more power as an organizational mediator on the ANMB.[7]

Meanwhile, Nelson and the WPB confronted increased pressure. Rubber and petroleum shortages, production bottlenecks, and work slowdowns forced Roosevelt to build new agencies that competed with and drained WPB authority. Soon there was a rubber czar, petroleum coordinator, and price stabilizer. Roosevelt gave Nelson power to coordinate, but the WPB chief dealt it away to army and navy administrators and then, pressed by his New Deal assistants, tried to get it back. His warmest liberal backers lost confidence in Nelson. "It appeared to me there had been a lamentable lack of overall planning," Henry Wallace worried. "I would say the fault is with the Joint Chiefs of Staff and Don Nelson."[8]

Baruch, Hopkins, and others urged Nelson to hire Eberstadt as a WPB vice-chairman in charge of developing a controlled materials plan. Nelson had nothing against Eberstadt but probably worried about any strong, driven mobilizers joining the WPB. "I have often thought, judging from my experience in the war program and earlier in NRA," Nelson wrote, "that quite a number of people would like to see a dictator dictating to the other fellow while they themselves remain free to operate as they wish." Nevertheless, he needed Eberstadt and finally offered him the vice-chairmanship in September, 1942. Now, it was the ANMB boss's turn to hesitate. For a number of administrative and intensely personal reasons, Eberstadt no longer trusted or respected Nelson. For one, Eberstadt believed that the WPB executive engaged in immoral social activities. However, Forrestal, Patterson, Somervell, and his "dear friend Admiral Robinson," chief of the navy's Office of Procurement and Material, urged Eberstadt to accept the WPB post.[9]

Baruch, too, pressed his mobilization disciple to direct the WPB requirements program. Increasingly solicitous of the clever and able Eberstadt, Baruch thought that with Eberstadt as a top WPB executive, he could augment his own growing influence in the wartime bureaucracy. (Baruch had recently been appointed to head the Rubber Survey Committee.) Whether or not Baruch envisaged Eberstadt as his agent for removing Nelson and making way for Baruch to come in as WPB director, he encouraged his younger Wall Street comrade to accept Nelson's offer.[10]

Eberstadt chose to enter the WPB hierarchy in order to enlarge the work of the ANMB and to implement and administer the Controlled Materials Plan. *Time-Life* publisher Henry Luce believed that Eberstadt's decision marked a new stage in the mobilization process. With the vision of building an "American Century" never far from his mind, Luce expressed his approval to his good friend Eberstadt. "All who know you and all who have worried so long and deeply about the progress of the war effort," Luce wrote, "will be feeling much better today." Now, Eberstadt operated on the inside of an agency that for too long had been seen as a wartime instrument for perpetuating the destructive competitive and socialistic programs of the

New Deal. Eberstadt recognized his new and wider opportunity to facilitate corporatization, for the WPB provided a real coordinating mechanism. "As I see the principal function of the War Production Board," he told the Business Advisory Council (BAC), a group of corporate leaders organized by Averell Harriman, during the Hoover administration, "it is as an intermediary or, if you will, *arbiter* between Government, which is our big customer, and production, which is the supplier." Eberstadt explained to the sympathetic BAC members gathered in November, 1942, that as WPB program vice-chairman he expected to influence organization. He planned to build a mechanism for a national economic program and to prevent wasteful competition for resources, labor, and manufactured products. The Controlled Materials Plan would become that interjacent mechanism between business and government long advocated by the BAC.[11]

In constructing the CMP, Eberstadt synthesized the ideas of organizational experts, former teachers, and professional associates. The art of modern administration, he insisted, lay in "a staff of competent people with whom you can exchange views preparatory to setting objectives." That made Baruch and President Hoover great leaders, Eberstadt believed, because as a starting point for policy they could "pick the brains of people whose opinion they value." Eberstadt had applied this method in his ANMB study and now expanded it for the far more significant task of developing a guiding mechanism for the entire war mobilization program. He recognized future implications for the postwar period as well, where a peacetime CMP could become part of a corporatist order. The idea that the war crisis gave those on the Good Man List a chance to revive progressive management and New Era associational innovations that had been interrupted by the Depression and New Deals always lay below the surface as Eberstadt began to assemble the CMP.[12]

Eberstadt found models for the CMP blueprint in German trade associations and in the liberal capitalist structures embodied in the Dawes and Young reparations commissions. The latter gave Eberstadt a model on which to base the structure and practice of the Dawes Transfer Committee, whereby a board of economic experts determined supply and demand. Eberstadt followed that structural innovation in setting up the WPB's Requirements Committee, which balanced the available supply of materials with the demands of army, navy, aircraft scheduling, Lend-Lease, the Board of Economic Warfare, and other claimant agencies. He also borrowed management innovations from GM and other giant corporations. He grafted onto the CMP a warrant plan developed by Cmdr. John D. Small and Maj. James Boyd, his military and naval assistants on the ANMB. Under this plan, the board issued warrants to manufacture based on availability of materials. The manufacturer became responsible for bringing production in line with materials.

Eberstadt also reviewed army industrial mobilization plans. He studied the British vertical steel allocation control plan introduced by Arthur Notman, a mining engineer and Baruch intimate, and WPB planner Charles Hitch, both members of the W. A. Harriman Mission for Economic Affairs to London the previous year. Eberstadt picked the brains of Courtney Johnson, Studebaker executive and classmate, as well as WPB experts Ernest Kanzler, Julius A. Krug, and David Novick. He relied on economist Bertrand Fox to pull the details together. "If I ever saw a team job, that was it," Eberstadt told Fox. "I just happened to be sitting in the driver's seat, where my tenure turned out to be rather limited."[13]

The Eberstadt team (most members of which now appeared on his wartime Good Man List) completed the CMP outline in late October, 1942, and circulated it to representatives of the major wartime contractors. Charles Hitch explained the plan to the National Industrial Council: "It is the basic mechanism for integrating strategy, production, and the flow of materials. It is the basic device for regulating the economy of America at war." Under this device, the WPB program vice-chairman collected data on available raw materials, estimated needs from each claimant agency, and gave the JCS an inventory of available materials for their strategic planning. Based on this information, the program coordinator drew up an overall schedule and divided resources among claimant agencies. These claimants then determined the best method to meet production goals and schedules in accordance with the predetermined supply of materials. The CMP created central control and decentralized operation, balanced supply and demand, eliminated competition for raw materials, and correlated forecasting with final production. Theoretically, it was an ideal corporatist mechanism. It created what David Novick called a "middle way" between unrestrained competition and complete state control of the economy.[14]

Once finished, the CMP had to be sold to suspicious New Deal administrators, who distrusted both Eberstadt and his plan. Eberstadt's outstanding characteristics, a confidential Bureau of the Budget memorandum noted, were his eagerness to please everybody and his "lust for power." At the same time, small businessmen favored a priorities system that allowed an individual scramble for materials and contracts. Proponents of a priorities requirements plan viewed the CMP not as a democratic compromise between socialism and fascism but as a vehicle for military-industrial domination of the economy. They feared the destruction of small business and saw the plan as a first stage for cartelization. They preferred the old priorities plan because it protected the interests of all producers, large and small, and every consumer, civilian and military, domestic and foreign. The new design promoted corporatist, noncompetitive industrial combination and military rather than civilian production. It was part of a desperate search "for some safe middle ground between Hitler and Abraham Lincoln," averred one critic who lamented that "there isn't any such place."[15]

Don Nelson feared the CMP, saying it reduced small business "to a condition of semi-feudal dependence upon big and middle-sized business, because of the necessity for subcontracting." The WPB chairman noted that already in 1942, the one hundred largest corporations held 80 percent of the defense contracts. He worried that the CMP's postwar impact would "not be wholesome," since it would crush small units and allow "acquisition of too much power by big business." In the end, Nelson accepted the CMP only because it promised to rationalize the chaotic wartime production process.[16]

Eberstadt calmed Nelson's fear. In a speech before the suspicious National Association of Manufacturers he presented the CMP as a patriotic blueprint to win the war. It was a mechanism for full war production, he explained, and for a "sound and healthy civilian economy." Eberstadt introduced the CMP as a guide for smaller manufacturers, not as an engine for their destruction. The plan sought only to get each industry its fair share of limited and scarce materials. "From there on, until the tank or plane or gun or piece of farm equipment rolls off the line," Eberstadt concluded, "the job of distributing material is up to you manufacturers." The CMP "just had to be adopted," Eberstadt's boss admitted finally, "in order to produce war goods."[17]

While endorsing the CMP principle, Nelson curbed Eberstadt's employment of the plan as a vehicle for larger postwar political-economic reorganization. With the president's direct intervention, Nelson recruited General Electric executive Charles E. Wilson as the vice-chairman in charge of production, leaving Eberstadt with responsibility only over programs and planning. "I had known Charley Wilson for a long time, and knew him to be a fair, able, and exceedingly forceful person," Nelson noted. Wilson seemed just the type to offset Eberstadt's influence. As expected, Eberstadt and Wilson battled for power and authority over WPB's industrial divisions, scheduling, and programming—the mechanisms for administration of the CMP. "The situation became more and more strained," Nelson claimed, "and developed into a conflict between Wilson and Eberstadt themselves, with Eberstadt having Army backing." Eberstadt saw it differently. There was no row with Wilson, he countered. He viewed Wilson as a product of the tough corporate politics of General Electric—"Detroit-Schenectady hard guy stuff"—who gravitated naturally toward acquiring authority in WPB. Eberstadt never blamed Wilson but admitted later that he should have gone to Forrestal, Patterson, Hopkins, or Corcoran as allies to quash Wilson's assumption of power in WPB. He didn't because he hated political maneuvering, Eberstadt insisted. "I didn't give a damn; I held it in contempt," he said.[18]

Despite his disclaimer, Eberstadt employed his own bureaucratic maneuvering to preserve and expand authority in the War Production Board. Of all the "tough guys at the top of WPB," David Lilienthal observed in late 1942, Eberstadt was the "'strong man' of the war production." Eberstadt practiced the current Washington definition of a coordinator as "a man who can keep all the balls in the air without losing his own." He established a Facilities Bureau under Fred Searls, Jr., a Baruch ally and mining executive, for control over construction planning. He hired classmate Thomas Armstrong to head the Foreign Requirements Liaison Branch, which had influence over Lend-Lease and the Board of Economic Warfare. He asked union leader Walter Reuther to bring more labor representatives into the WPB industrial divisions. He collaborated closely with the War Department and placed military and naval representatives from the ANMB in close partnership with their counterparts on the WPB, a move that was heavily criticized as a step toward military control. Eberstadt fought every attempt to drain his authority. When Leland Olds of the Power Commission, backed by Secretary of the Interior Harold Ickes, requested that the WPB turn over its data on electrical supplies to his office, Eberstadt warned Patterson that Olds sought to gain more influence. "This looks innocent but it's a crucial entering wedge in the Commission's campaign for control [over all electric power]," Eberstadt said. "Whoever determines supply and requirements," Eberstadt reminded his friend, "has the whip hand."[19]

In early 1943, it remained unclear who determined supply or requirements: Eberstadt or Wilson, the War Department or the WPB. Instead of solving management problems, the recruitment of Eberstadt and Wilson and development of a controlled materials plan sharpened tensions and power struggles. The military complained incessantly about inadequate supplies. Material bottlenecks and production delays continued. Criticism from both New Dealers and the more conservative war mobilizers focused on Nelson's lack of leadership and poor management skills. Liberals blamed Nelson for losing control over the army. Wallace warned the president that men in uniform and their business allies on the WPB wanted big corporations to run the country for the next ten years. Liberal jurist and FDR confidant Felix Frankfurter agreed that the "ruption [*sic*] inside of the WPB" meant that Nelson could not control Eberstadt, "who with characteristic ruthlessness and disregard of the ordinary canons of loyalty has for long been working with and through the services" to advance the interests of big business.[20]

On the other hand, Nelson's mismanagement was accused of leading to industrial chaos, wasteful competition, profiteering, and shortages. "The condition of the WPB and the war production organizations," Eberstadt observed, "expressed a political, labor, and capital division existing in the country prior to our entrance in the war." If Nelson could not discipline individual interests and prevent conflicts, perhaps the Good Men in business and government could. And once the war ended, they wanted to continue that leadership. It was left to the business leaders, Eberstadt wrote in January, 1943, to develop plans "with a view to seeing to it that after the war there shall continue to be steady employment for all accompanied by such business and governmental policies as will permit a steadily rising standard of living in this country and abroad."[21]

Tensions increased. "The question then was: who is to be the Number Two Man in WPB–Eberstadt or Wilson?" one WPB official asked. Having lost authority to Wilson and being convinced that he had done all that he could to develop the CMP, Eberstadt decided in early February, 1943, to resign his vice-chairmanship. He went to Baruch with the news. The aging mobilization expert asked Eberstadt to wait, explaining that he would give his reason when he returned from a quick trip to New York City. Baruch failed to reveal his motive to Eberstadt at the moment. However, that same day the president asked Baruch to take over the struggling production board. Baruch wanted as his two WPB directors Eberstadt and old WIB assistant John Hancock, a Lehman Brothers investment banker who sat on the board of directors of nineteen major corporations. Baruch went to New York to ask Hancock to join the WPB team. But Hancock was unavailable due to an illness in his family. Suddenly Baruch took to bed in New York complaining of fatigue, while the controversy surrounding the War Production Board raged on in Washington. Thus Baruch stood above the apparent plotting to overthrow Nelson and to establish corporate management of the mobilization program.[22]

As he waited for Baruch's return, Eberstadt took the first break from government service since Pearl Harbor day to visit his son at Exeter Academy. During the trip, he picked up the *New York Times,* reading the headline: "Eberstadt to Go Washington Hears." Eberstadt raced back to Nelson's office for an explanation, where the WPB chief denied either leaking such information or planning to remove his assistant. Nelson begged the vice-chairman to stay with the board. Two days later Nelson fired Eberstadt. "He did this so fast that the announcement was ready for distribution in the OWI [Office of War Information] press room before Nelson got around to telling Eberstadt about it," recalled Bruce Catton, who was a WPB public relations officer.[23]

The details of and motivation for this critical bureaucratic decision remained clouded in controversy. To Don Nelson, it appeared a simple administrative move. "I did my best, but without success to resolve this conflict," he insisted. However, Eberstadt's methods and philosophy of advocating army control over the CMP did not coincide with WPB policy. It was a question of civilian versus military priorities, Nelson concluded. Others saw the firing from a different perspective. Bruce Catton believed that the White House put pressure on Nelson to remove Eberstadt. *Time* publisher Henry Luce propounded the view, shared partly by his friend Eberstadt, that a small clique of New Deal administrators and economists on the board persuaded Nelson to get rid of Eberstadt. More likely, Harold Smith and his assistant budget director, Wayne Coy, privy to the president's intention of hiring Baruch, informed Nelson of the attempt to replace him. Smith hated Baruch's kibitzing and undoubtedly knew about the budget bureau memorandum warning of Eberstadt's "lust for power." Smith feared the two corporatists' employment of a dynamic WPB machine to bypass executive authority and presidential power. Apparently, Smith sent Coy to visit Nelson and told him to save his job by eliminating Eberstadt.[24]

Eberstadt held his own theory. He thought that a vengeful Roosevelt personally orchestrated his removal from the war mobilization agency. "The President, as it happened, was not fond of Eberstadt," Catton recalled. A private meeting between Eberstadt and Roosevelt in October, 1942, left a bitter legacy. As Eberstadt explained it, the president asked him to fire Tom Armstrong. Roosevelt admitted that Eberstadt's Princeton classmate had done nothing wrong, only that for "reasons of state" he must leave the WPB. Eberstadt suspected that an old feud between Thomas Armstrong, a Standard Oil executive, and Assistant Secretary of State Sumner Welles over oil concessions in Mexico prompted the president's request. Moreover, FDR appeared sensitive to recent liberal press criticisms of Armstrong's alleged conflict of interest on the board's foreign distribution division. "Too bad about Armstrong," H. B. Swope wrote Eberstadt, "but I see no other way out than that he go."[25]

Instead, Eberstadt defied Roosevelt. He refused to fire Armstrong, who had given up a top management position with the Rockefellers in order to assist his friend on the WPB. In addition to being a fellow member of the Class of 1913, Armstrong and Eberstadt had been together on the Princeton Literary Club and in the Plattsburg reserve officer program. Under no circumstances would Eberstadt betray this ideal Good Man, who had been on the list since the 1920s. Finally, Eberstadt convinced the president to leave Armstrong alone until the CMP began to function, and then both dollar-a-year executives would return to private business. Roosevelt dropped the matter for the moment, but Eberstadt believed that FDR waited for an opportunity to fire him. In fact, Eberstadt overestimated the president's personal interest in the case. FDR wanted the WPB to function more smoothly, preferably under Nelson, but appeared ready to accept the advice of close advisors Jimmy Byrnes, Harry Hopkins, and the service secretaries to promote Baruch. Roosevelt had accepted the elevation of Eberstadt, but with him out of the picture Roosevelt would make no effort to regain his services.[26]

Whatever lay behind this sensational firing of a mobilization official during the war, Eberstadt felt betrayed. By asking him not to resign, Baruch made Eberstadt a scapegoat in the mobilization power struggle. "He was my friend and I don't resent it," Eberstadt insisted twenty-five years later, "but maybe he shouldn't have asked me to unless he was sure that he was going to take it." At least Forrestal and Patterson fought "bitterly" and "sharply" with Nelson to save Eberstadt's job, but they failed as well. Eberstadt brooded. In his own mind, he served as a nonpartisan expert and arbiter between government and business. He thought himself above political intriguing, so forced removal came as a bitter blow. After being pulled from a meeting and told that Nelson had dismissed him, Eberstadt was so upset that he rushed from the building, leaving behind topcoat, hat, and personal belongings. He told friends that he would never take another Washington job. He went to Stimson and asked for active duty in the army. The secretary of war said Eberstadt was too old for combat service and instead offered a desk job. Eberstadt refused it.[27]

The depth of Eberstadt's feelings after his removal from the WPB appeared in a number of unusually virulent comments made in early 1943. Though these were intensely personal observations, they reflected a growing anti-Roosevelt sentiment and wider wartime reaction to New Deal leadership, policy, and organization. Eberstadt attacked Nelson as "contemptible, cowardly, and scheming." He then criticized Henry Wallace, the inspirational leader of the liberal faction on the WPB, once again voicing not only his own perception but the opinion generally held by dollar-a-year men and political opponents. "Wallace [was] both a dreamer and a sleeper," Eberstadt observed. "He used to have most of his dreams during meetings of the War Production Board." Worse, Roosevelt had become, in Eberstadt's words, "an apostle of confusion" who unleashed fools and misguided visionaries to run the mobilization program. He blamed Roosevelt for the lack of coherent and consistent policy, for vacillation, and for "stupidity" in surrounding American goals, organizations, and plans with a perpetual "air of mystery." The president created "humpty-dumpty organization" that perverted the WPB into a "cesspool of intrigue," rather than making it the modern, efficient corporate agency necessary to coordinate production. FDR diverted attention from his own poor leadership and faulty organization by working through Senator Truman's investigating committee, which Eberstadt called "a new high in buckpassing."[28]

Personal rancor and deeper underlying philosophical differences drove

Eberstadt toward that amorphous aggregation identified as the "we-hate Roosevelt people." This group included Henry Luce and Clare B. Luce, Sen. Kenneth D. McKellar, and industrialist Thomas J. Wallner of the Southern States Industrial Council. Eberstadt joined the chorus, claiming that the president was playing a soldier-politician who bred selfish interests and promoted bureaucratic infighting. During the war crisis, FDR perpetuated poorly conceived New Deal agencies that had failed to solve the deep economic and political problems revealed by the Depression. The WPB embodied these New Deal flaws. More ominously for Eberstadt, New Dealers planned to employ the WPB "to control business after the war." Nelson circulated a WPB plan that incorporated the regulatory prescriptions of Henry Wallace, Robert Nathan, and other liberals. Eberstadt felt that they overregulated, pandered to factions, formed competing offices, and stimulated conditions that fostered political extremism and violent economic cycles of depression and inflation at home and abroad. "Strikes, race riots, etc. are due to growing contempt for the government," Eberstadt observed. New Dealers raised phony ideological confrontations, he continued. Every time they wanted to seize another function, Eberstadt lectured Patterson, they screamed "military takeover."[29]

Eberstadt dismissed ideology as a smoke screen that obscured power struggles between loosely defined wartime mobilization groups. These groups differed on organizational and political-economic principles, but all believed in the preservation and expansion of the liberal, capitalist system. Eberstadt never shared Forrestal's fear of Marxist ideology as the most dangerous threat to American interests in the postwar world. Eberstadt did find communists in the New Deal bureaucracy but faulted them more for incompetence and weakness in fighting the common enemy than for their ideology. Eberstadt also did not frame the struggle between New Dealers and his Good Men as a fight between liberalism and conservatism. He saw it as a conflict between naive visionaries and objective professionals, who were, he hoped, "liberals but not New Dealers."[30]

Nevertheless, ideologues such as Forrestal and Patterson gathered about Eberstadt. Despite his announcement that he was through with the Washington scene, Eberstadt stayed close to his friends. Ironically, by removing Eberstadt from his government post Nelson and Smith made him an even greater threat to FDR and New Deal administrators. Now, Eberstadt wandered freely between departments and agencies, advising those that he considered the Good Men in government and spreading his basic message that "the necessity for a comprehensive reorganization is indicated." He prepared studies and memoranda on demobilization, labor resources, postwar plans, and reconversion for Patterson, Forrestal, and anti–New Deal legislators such as Congresswoman Clare B. Luce. He discussed the development of a noncompetitive postwar commercial aviation syndicate with Robert Lovett, assistant secretary of war for air. He encouraged Julius A. Krug and other WPB allies to develop the CMP for wartime *and* postwar reorganization. He joined Baruch's team of "kibitzers" as they searched during those critical last months of war for a corporatist political-economic blueprint to determine the course of postwar American economic, military, and foreign policies. He worked behind the scenes to organize opposition to FDR's reelection in 1944 and became a facilitator between anti–New Deal forces in the private sector and the Good Men in government. He wanted to "unite the like-thinking elements of both parties," he told Chamberlain.[31]

Nelson worried about Eberstadt's continued presence in wartime Washington. After he fired the WPB vice-chairman, Nelson reorganized the board and worked on a reconversion plan that gave WPB primary responsibility for the resumption of civilian production in order to protect small business from the larger corporations that dominated war contracting and production. Nelson's reconversion plan drew immediate opposition from the War Department, which argued for continued all-out war production. This approach prevented small business from getting a jump on peacetime reconversion. GE executive Charles Wilson sided with the War Department position, and soon the WPB divided again into two factions. It was another stage in the struggle for a political-economic system that transcended the immediate issues of military shortage or production and focused on who would control postwar economic policies. Nelson suspected that Eberstadt was deeply involved, believing him to be the leader of an anti-Nelson conspiracy that included Forrestal, Patterson, Baruch, General Somervell, and rubber czar William Jeffers. Eberstadt appeared to be behind the War Department's attempt to employ the CMP as its own vehicle and to block WPB control of the CMP. "To speak frankly," Nelson wrote, "the Army could at any time sabotage any program I set up."[32]

With Nelson placed in this context, he possibly ordered a close watch kept on Eberstadt's activities. Baruch told Eberstadt that Nelson spied on him. "He thought that the FBI had been watching all of my acts and movements with great care," Eberstadt observed. "He told me that he had very good information that my telephone conversations were being listened in to, that possibly there was a Dictaphone in my house." Simultaneously, the Luces informed Eberstadt that Nelson had hired "three whisperers" to defame his character and had planted rumors that he had stolen a key government document.[33]

The situation had become so disgusting, Forrestal told Eberstadt in early July, 1943, that he was thinking about resigning as undersecretary of the navy. Forrestal never intended to quit his critical mobilization job, but the threat delayed Eberstadt's departure from Washington. Moreover, it reflected the chemistry of the Eberstadt-Forrestal partnership. Every time that Forrestal felt pressure building, he raised the resignation issue with Eberstadt so that his friend would offer reassurance. Eberstadt never disappointed him. This time, he insisted that Forrestal was indispensable to the war effort. More important, he told Forrestal that he must stay and direct the "clean up job after the war."[34]

No matter what Forrestal's motive was, the threat to resign revealed the intense intriguing that convulsed the WPB and mobilization machinery in mid-1943. The president recognized, too, that the controversy swirling about Nelson and the WPB interfered with war production, the conduct of smooth industrial relations with government, and not incidentally, his administration's public image. On May 29, Roosevelt appointed Democrat James F. Byrnes, a former U.S. senator from South Carolina and Supreme Court justice, as the director of the Office of War Mobilization (OWM). "It looks like some more of the old Roosevelt prescription, an additional top story on a building whose foundation is rocking," Eberstadt worried. He moderated this view when Byrnes hired mining engineer Fred Searls, a "crusty gentleman of the Republican persuasion" and close Baruch associate, to organize the OWM. The only executive on Eberstadt's Good Man List to resign his WPB post in protest of Eberstadt's firing, Searls sought Eberstadt's advice on the staffing and organization of OWM. Moreover, Byrnes appeared to have the right corporate instincts. "He is a decent conservative, strongly influenced by Baruch, Hancock, etc.," Lilienthal noted. Byrnes set up a committee of expert advisors that included Baruch, Stimson, and Frank Knox. He wanted Baruch and Hancock to prepare studies for OWM on demobilization, reconversion, labor, and industry. In turn, Baruch asked Eberstadt to develop a political-economic framework for these reports on postwar economic issues.[35]

Baruch's request crystallized Eberstadt's emerging political-economic synthesis for postwar American economic stability and national security. Throughout the early autumn of 1943 Eberstadt further refined the concept of the CMP as the coordinating mechanism for American industrial democracy. He advised Forrestal that the CMP provided a "perfectly simple mechanism" for solving labor problems that plagued naval production, particularly the severe shortage at West Coast aircraft plants. Like material shortage, Eberstadt explained, labor scarcity hinged on that fundamental equation of supply and demand and was subject to the same control. Once a central government agency determined national needs, defined production goals, and allocated materials, then industrial divisions contracted with those producers who had the labor supply necessary to complete the contract. If business wanted contracts, Eberstadt advised Forrestal, they would pay the wages to attract the needed workers.[36]

Eberstadt expanded the CMP blueprint in a twenty-one-page study that he entitled "The Principles, Procedures, and Organization of Industrial Demobilization" and submitted to Baruch and Hancock in November. Eberstadt sought in this essay "to advance political and economic theories," and to define a mechanism for "winning the peace" through "the leveling off of peaks and valleys within and between various industries." He thought that this could be accomplished by reversing the original mobilization process. In demobilization the CMP provided a mechanism for the transition to peace by preventing inflation, overproduction, and hoarding by individual interests (or small business) trying to rush into peacetime production. "Wherever studies indicated that postwar demand will result in serious shortages of raw materials," he told the OWM advisors, "material control should be maintained with a view to an equitable distribution of the critical materials amongst the several competing industries and amongst the several companies within such industries, with particular regard to protection of the small user."[37]

Eberstadt hoped that Baruch had enough influence with James Byrnes to employ the OWM as the central coordinating body with which to implement Eberstadt's CMP blueprint. He urged Baruch to organize such a demobilization program around four ideally noncompetitive functional groups: military, industry, labor, and Congress. These sectors–not publicly elected officials–should determine postwar policy. The OWM had to provide interorganizational coordination. The time to plan was now, Eberstadt concluded, so as to avoid postwar political and economic chaos that lay at the root of war.[38]

Eberstadt believed that such corporate planning was under way. Baruch and Hancock applied much of his memorandum to their plan for Byrnes, who had emerged as the central mobilization coordinator. Meanwhile, WPB struggles led to Nelson's departure and diminished WPB influence over reconversion. WPB chief J. A. Krug worked with Byrnes and the War Department to slow civilian reconversion, in effect ensuring that the larger businesses would regain their prewar dominance once peace returned. Byrnes seemed to side with the War Department's view of a reconversion timetable favored by big business and by many of the dollar-a-year mobilizers in wartime government. Other positive signs had Eberstadt convinced that a corporate design might prevail in reconversion planning. Assistant Secretary of Commerce William Clayton, instead of the more liberal Lilienthal, became surplus property administrator. Control over the distribution of surplus property, Eberstadt knew, meant postwar control by surplus capital held by the largest corporate interests.[39]

However, there was no consistent approach to reconversion. Byrnes adopted bits and pieces of the Baruch-Hancock reconversion design, and as Baruch biographer Jordan A. Schwarz has argued, blocked Baruch's efforts to make the OWM "into the corporate architect of the postwar economy." Byrnes turned reconversion policy and planning over to the White House and Congress, where New Dealers, budget planners, and WPB allies such as Senator Truman felt that it belonged. Instead of the corporatist path favored by Eberstadt, Baruch, and Hancock, reconversion planning went in various directions and levels from proposals that advocated extensive government regulation and control of corporations to a more associational approach that left reconversion authority with corporate dollar-a-year men in their war agencies. Undersecretary Patterson explained the situation to Eberstadt. "The other side got off to a flying start, and could not be headed off," he complained. "First it was the 'WPB spokesman' line, then everybody charged in."[40]

None of these reconversion efforts–with the possible exception of the Baruch-Hancock plan–satisfied Eberstadt. None of the plans for postwar reconversion presented a clear vision of political-economic policy and organization. They all lacked what he called a broad-gauge understanding of postwar problems. They recommended no structures with which to link business and government or foreign and military policy. Eberstadt lost faith in Byrnes, who seemed to ignore Good Man advice. "Where are the planners' plans and blueprinters' blueprints?" Eberstadt lamented. He felt that the chaotic postwar planning left business, labor, and the public confused and uncertain about reconversion. "The sad part," Eberstadt warned Congresswoman Luce, "is that unless we can have a sound reconversion policy, not only are we going to suffer domestically, but our foreign policy, which is based upon a healthy domestic condition, will go by the boards as well."[41]

Eberstadt could not have known at this stage that in a few months Forrestal would ask him to prepare a report on postwar national defense reorganization that would provide further opportunity to develop his own design for postwar organization. Nevertheless, Eberstadt prepared the rough outline of a postwar system that linked political-economic organization to a larger concept of national security. Eberstadt aired his program in a speech that he delivered at Princeton University, where he received the Class of 1913 Cup given in recognition of his wartime service to the nation. He had his speech printed in a pamphlet and distributed to those on his Good Man List. His speech thus joined the dozens of other plans for postwar America then circulating about Washington.

Eberstadt drew inspiration for his university talk from the ideas outlined by close friends Henry Luce and John Foster Dulles in their respective essays "The American Century" and "Six Pillars of Peace." Echoing his friends' writings, Eberstadt announced that their generation had the opportunity–perhaps the last one–to control the epidemics of war and unemployment that had disrupted the world for the past quarter century. Wilson's generation had been given a similar golden opportunity after the first world war to develop a century of American world leadership but had stumbled badly. This time, the men who organized American mobilization between 1940 and 1944 had the means to assert world leadership and promote this "American century." "The war has taught us how we can use the miracle of full production" Eberstadt explained.[42]

Eberstadt expressed concern in his Princeton speech that the Roosevelt government, which he called privately a scandal of inefficiency and ineptitude, could not provide postwar leadership for constructive business-government cooperation. "Government should be the guardian and protector of legitimate business," he told the sympathetic Princeton audience. "It should reconcile class and sectional differences," he said, and it should reveal clear postwar plans so that business could make the transition to peacetime production. Government needed to develop tax, banking, and trade policies that removed unnecessary regulations and artificial hindrances to wealth. Drawing on his experiences in preparing the Dawes Investment Brochure and his membership on the Young Commission, Eberstadt called for "prudent public and private loans" to rebuild war-shattered nations, which would create healthy customers and areas for new investment and commerce.[43]

By raising the world's standard of living, Eberstadt continued, the American business-government partnership contributed directly to international stability and to what his friend John Foster Dulles called a just and durable peace. Eberstadt revived the views of his old law firm mentor Joseph Cotton and of Dwight Morrow and Herbert Hoover, who had called for the development of an international food policy to rebuild war-ravaged countries. Economic and moral leadership alone would not create this "American century." Eberstadt introduced similar ideas in his ANMB and CMP studies, but for the first time he now tied America's economic organization and defensive might directly to an emerging blueprint for postwar national security. "This is the keystone of the structure of the achievement and maintenance of a just and durable peace," he concluded before the appreciative Princeton gathering.[44]

Eberstadt's evolving design for postwar American economic well-being and international security approached what historian Thomas J. McCormick has identified as a system of globalized corporatism. It forecast, too, the international projection of American corporatist neo-capitalism to build a postwar economic order at home and abroad that historian Michael J. Hogan found at the base of the Marshall Plan. Moreover, Eberstadt's views reflected a rising conservative or neo-realist reaction to Wilsonian idealism and New Deal liberalism generated by the nature of the world war and the intensely competing ideologies that had manifested themselves as war. Eberstadt's speeches and writings during 1943 and 1944 translated what contemporaries labeled the neo-realist perspective of Henry Luce, John Foster Dulles, Walter Lippmann, Herbert Hoover, Hugh Gibson, and Reinhold Niebuhr, among others. Eberstadt joined the conservative theologian Niebuhr, along with Baruch and Dulles, in contributing essays to the Christian Council of Churches publication *Postwar World*. The widow of George S. Franklin, Eberstadt's former law partner, edited this conservative newsletter, and she relied on Eberstadt to help spread the council's message of international economic development and world order. She wanted Eberstadt to distribute material provided unofficially by her friend in the State Department, Harry C. Hawkins, director of the Office of Economic Affairs. She wrote Eberstadt that she would introduce him to Hawkins, adding, "I know you have so many connections down there that probably this would not be necessary."[45]

In reality, Eberstadt's only significant Washington connection remained Forrestal. Their public-private partnership had brought Eberstadt to Washington in the first place to sort out the machine tool and personnel recruitment problems. It had given Eberstadt an opportunity to develop his ideas on political economy and organizational structures for the ANMB and Controlled Materials studies. His connection with Forrestal allowed him to implement some of these ideas when he served as ANMB chairman and WPB vice-chairman. The Forrestal connection continued to be Eberstadt's actual avenue of influence in the postwar reconversion process. He never coveted such influence, nor expected to play a significant role in postwar politics and reorganization. Eberstadt's correspondence with Forrestal, which included a draft of his Princeton speech, demonstrated his attempts to imbue Forrestal with corporatist principles and to persuade his friend to use defense reorganization as a vehicle for larger political-economic purposes.

Eberstadt bombarded Forrestal with partially articulated organizational and philosophical principles, elements of which made a deep impression on Forrestal. Eberstadt insisted that government had to develop a consistent and realistic national security policy. There was no room for emotionalism. Peace had to be planned by hard-headed realists such as those who had written the U.S. Constitution or had assembled the Congress of Vienna. The latter, Eberstadt explained again to Forrestal as he had in 1941, were "some very wise and experienced men, but pretty cold and unemotional fellows [who] laid the plans and guided its course." That same realism and balance-of-power philosophy ought to govern the peace after this war as well, in order to avoid the mistakes of 1919, when the victors sought revenge upon Germany and in effect created the destructive debts-reparations cycle. "The rational unemotional objective, as I see it," Eberstadt advised Forrestal, "is after having thoroughly horse-whipped those in Germany that were responsible for what has occurred and mostly for the way in which it was done, to attempt to bring 80 million hardworking and productive people back into the family of nations under circumstances where peace will be more palatable than war."[46]

Eberstadt reminded Forrestal that the United States must maintain its defensive might after the war. He found efforts at international organization, like those under consideration at the Dumbarton Oaks meeting, only a beginning and not a substitute for a firm national security policy. "As Dumbarton Oaks and other segments of the postwar world begin to unfold," Eberstadt wrote, "I feel more strongly than ever the importance of accompanying our sincere and vigorous efforts for a better world with strong military defenses and adequate preparedness against any possible slips 'twixt the cup and the lip.'"[47]

Throughout 1944 Eberstadt looked for a great American leader to shape policies for the postwar world. The United States needed leadership from those with wide political and economic knowledge, from those who possessed the qualities of statesmanship that Eberstadt perceived in James Madison, Alexander Hamilton, and George Washington. Eberstadt added that such statesmen ought to have the moral dimension of Protestant reformer Martin Luther. He doubted that the tired and worn Roosevelt was the great Christian statesman who could rebuild the world and design postwar structures for peace and security. Instead, Eberstadt predicted that Roosevelt would become another Lincoln or Wilson, who were removed from postwar reconstruction processes. He hoped that Dewey might provide the necessary leadership, and Eberstadt worked hard on Dewey's presidential campaign, advising and rewriting speeches for the Republican candidate.[48]

With Dewey's defeat, Eberstadt worried that even if Roosevelt survived, the president would follow the advice of that sleepy visionary Henry Wallace and plan for a "Century of the Common Man" rather than Henry Luce's vigorous "American century," or opt for Sumner Welles's hazy, idealistic "Blue-Print for Peace" rather than John Foster Dulles's more concrete "Six Pillars of Peace." Most important, Eberstadt feared that Roosevelt would follow treasury secretary Henry Morgenthau's design to dismember Germany rather than the Good Man vision of a strong, economically sound Germany as the stabilizing factor in postwar Europe.[49]

Eberstadt insisted that Forrestal and Patterson assume a larger role in reconversion, since they both understood the political-economic concepts behind the Controlled Materials Plan. They also recognized the importance of the ANMB for easing the transition from war mobilization to peacetime production and shared the Eberstadt view that corporate organizers (especially those on the Good Man List) must shape the course of postwar national policy and organization. Eberstadt urged Forrestal to join Patterson in employing their departments as key agencies in the demobilization process that would redirect labor, facilities, and resources to peacetime uses. "It may be that here too in the end," he told Forrestal, "the services will have to take the ball and *force* a clear and sensible program of demobilization just as in the old days."[50]

Eberstadt was right. Forrestal, Patterson, and the armed services played pivotal roles in the politics of reconversion. Their participation arose as a result of the army's quiet introduction in late 1943 of a plan for a single department of national defense. This proposal had blossomed by 1944 into an open debate over what was referred to as defense unification. By 1945, it had become a full-scale power struggle. This "unification" phase of reconversion continued struggles between New Dealers and dollar-a-year men over larger political-economic designs for America begun during the mobilization process and intensified during the war. Military issues and interservice fighting obscured the connection between mobilization and unification and between various groups involved in both. In the end the central roles played by Forrestal and Eberstadt assured a direct institutional linkage.

5. Reconversion and Unification, 1944–1945

The reconversion debate grew more intense throughout early 1944. Everyone had a plan—from the Truman Committee to the State Department, from the WPB to Baruch and Hancock, and from the Bureau of the Budget to the Treasury Department. Even the army's chief of staff, George Marshall, introduced a secret plan on postwar governmental reorganization to the Joint Chiefs of Staff. "There were reports, conferences, committees and subcommittees to study, to make findings, to coordinate, and to activate, and there were special orders and directives," a WPB official recalled. These plans represented two designs that echoed the positions taken during the mobilization debate by New Dealers and their congressional allies such as Truman on the one side and the dollar-a-year experts on the other. The first plan called for early reconversion to civilian production so that the government could assist small business and organized labor, while the other side held that full war production must continue until the end of the fighting, a position sure to benefit and preserve the great corporate interests and big business that dominated the economic scene before the war.[1]

The reconversion debate included as an important adjunct a discussion of the institutional changes and structural innovations implemented during mobilization and war. The forms of postwar organization reflected (and helped determine) the direction of the reconversion debate. Would temporary mobilization structures—what Senator Truman called "the various extra curricular agencies" such as the WPB, Joint Chiefs of Staff (JCS), or Office of Strategic Services (OSS)—continue in some permanent form after the war? If so, would they be absorbed into regular government departments or become new coordinating agencies? Would the federal government, particularly the office of the president, continue to grow in power and authority? Or, would a more associative form of government organized around public and private coordinating structures (somewhat as envisaged by Herbert Hoover and his New Era Republicans before the Great Depression) reemerge to provide corporatist mechanisms to administer postwar America?[2]

One of the more obscure plans that emerged among the many designs for postwar organization provided the catalyst for the reconversion debate during the last months of World War II. This was Gen. George C. Marshall's plan for a "Single Department of War in the Post-War Period," submitted to the JCS in late 1943. The ranks of single-department opponents and advocates formed along roughly the same lines that developed in the fight over mobilization and war production. In general, New Dealers favored a single, unified national defense department, while the dollar-a-year organizers favored a system of coordinating agencies for national defense and security.

The War Department, a vigorous ally of the dollar-a-year forces during mobilization, war production debates, and early reconversion opposition, switched sides dramatically during the single-department controversy. A powerful bond between General Marshall and former National Guard officer Harry Truman and between the Bureau of the Budget and War Department bureaucracy contributed to this switch. Consequently, this aspect of reconversion was far less a civilian-military conflict, as many contended at the time, than a fundamental power struggle between groups of civilians with different organizational and political-economic visions of postwar America.[3]

When Marshall submitted his plan for a single department of war in late 1943, he never expected the bitter defense unification conflict that continued long after the creation of that single department of defense. Nor could he have predicted that his plan, which on the surface seemed simply a proposal to combine the War and Navy departments and create a single Cabinet post for defense and single military chief of staff, would become the catalyst for a far larger political-economic and organizational struggle.

Marshall initiated a critical phase of reconversion with his reorganization plan and became, in many ways, the Eberstadt-Forrestal team's main antagonist in the process of building a national security organization. Consequently, we need to diverge from Eberstadt and Forrestal in order to detail Marshall's organizational background, motivations, and conduct. Throughout his career, Marshall had adopted what he referred to as that "middle ground" that governed "almost every plan for the control of human affairs." He favored compromise, not confrontation, and cooperation rather than controversy. As army chief of staff since September, 1939, Marshall had made necessary compromises for interservice harmony. He instructed army officers not to "quibble" over minor matters with their naval counterparts. He developed good relations with Harold Stark, chief of naval operations. Marshall adjusted his own service by tying the increasingly independent-minded and militarily important army air force (AAF) to the War Department organization by making it one of three co-equal arms with the ground and supply services. He toiled against political interests and entrenched bureau opposition to unify military supply in a single agency under Gen. Brehon Somervell. In the JCS, formed in early 1942, Marshall cultivated harmonious relations with the navy's commander-in-chief, Adm. Ernest J. King, and the JCS chairman, Adm. William D. Leahy.[4]

Yet, two years later, Marshall departed from his middle ground to introduce the plan for a single department of war, which elicited immediate navy opposition and bitter controversy. From the outset his proposal threatened established practices and patterns of authority, tore open interdepartmental wounds, and resurrected old rivalries. Even before the Allies had established the Normandy beachhead or launched the final campaigns across the Pacific, the American defense bureaucracies began to feud over postwar organization, roles, missions, and policies. Congress, the budget bureau, press, and public jumped into the fray. The single-department plan accentuated growing legislative-executive conflict over postwar leadership and sharpened the struggles between the two competing power groups. It forced Forrestal to lead a countercharge against the Marshall scheme, which he began by hiring Eberstadt to develop an alternative plan.

Why did Marshall, the master of compromise, introduce a single-department scheme that included a military chief of staff, an idea sure to provoke anitmilitarist feelings and upset traditional organization? The answer lay in Marshall's frustrations during four decades of service with the separate army and navy structures and his views on universal military training for a citizen army. His single-department scheme brought him into the circle of those who favored greater presidential authority and federal control —that is, many New Deal bureaucrats, public administration experts, and not incidentally, Marshall's warmest admirer, Sen. Harry S. Truman.

The intensity of Marshall's pursuit of the single-department formula arose from his own military experience. The apparent inequity between army and navy caused by the two-department system had confronted Marshall as he entered the army in 1902 after graduation from Virginia Military Institute. His first duty took him to the Philippines, that new outpost of American imperial and naval power, where he joined a military establishment inadequately prepared and organized to carry out land operations overseas. "In those far off days," Marshall recalled in 1950, repeating a theme that bothered him throughout his career, "the soldiers of the regular army got little attention or consideration from the Government or the public."[5]

Vigorous rebuilding of the War Department between 1902 and 1903 under Secretary of War Elihu Root, who constructed an Army War College, standardized militia system, Army General Staff, and Joint Army Navy Board, failed to strengthen the army's position in relation to the navy. Subsequent administrations emasculated the general staff, misused the war college, and neglected the joint board. Theodore Roosevelt, who shared Root's larger cosmopolitan views and supported War Department reorganization, nevertheless neglected military development. The army had to remain a mere "skeleton," Roosevelt insisted, and must not cause the "slightest appreciable tax upon the taxpaying citizens of the Nation." A disciple of late nineteenth-century naval historian Alfred Thayer Mahan, Roosevelt favored the "upbuilding" of the fleet as "the lightest premium for insuring peace which this nation can possibly pay." His reliance on traditional naval organization as the first line of defense threatened to bury Root's innovations and relegate the army to inferior status in the emerging national defense organization.[6]

Such treatment of the army left permanent scars on young officers like Marshall who began their military careers during the Theodore Roosevelt administration. He found an unfavorable contrast between treatment of the sea service with that of the army in everything from navy domination of defense roles and missions to the most minuscule detail of service uniforms. "What I cannot understand is why the Navy and Marine Corps seem always to be able to do better in this clothing matter, year after year, ever since I have been in the Army," he wrote the quartermaster general. "I know they have more money because they are more popular with Congress and I believe they are less rigidly held to account for the details of their expenditures," he added.[7]

The disparity between army and navy peaked for Marshall during the 1930s, as Franklin Roosevelt, former assistant secretary of the navy and now president, used New Deal monies to rebuild the fleet, improve shipyards, and restore full ship complements cut during the Depression. Though he pursued a gradual approach to naval rearmament, FDR's policies seemed to Marshall to neglect the army, especially since he found not one tank available to train National Guard units. He complained that "unlike the Navy Department, the War Department administers the Army on a parsimonious basis." When aggressive army aviators agitated for a larger share of the defense budget, Marshall told the Air Corps Tactical School at Maxwell Field that the entire military suffered from an imbalance in resources. "Sit down sometime and try to balance all the factors concerned with the national defense, including limited appropriations, and then attempt to outline the organization for a balanced Army," he wrote.[8]

Marshall saw a remedy in a single department that would oversee both army and navy and in a system of universal military training (UMT). These aspects of Marshall's thinking evolved simultaneously, were interrelated, and owed substantial influence to close friend John McAuley Palmer. Marshall first met Palmer at the Fort Leavenworth command school, beginning a lifelong friendship from what they called their Leavenworth days of Army reformation. Over the years the more scholarly West Point graduate Palmer taught Marshall military history, the UMT principle, and the importance of the single department of defense. During the first world war and the immediate postwar era they served on Gen. John J. Pershing's staff, where they promoted UMT and a "Ministry of National Defense." Palmer collaborated with James Wadsworth, chairman of the Senate Military Affairs Committee, on writing the National Defense Act of 1920, while Marshall testified for a single department of defense before President Harding's Joint Committee on the Reorganization of the Administrative Branch of the Government.[9]

When Palmer resigned from the army in 1926, his influence on Marshall increased. Marshall reviewed chapters from Palmer's manuscripts on military history, promoting the retired officer as "one of the best brains in the Army." Palmer's study entitled *Statesmanship or War* most impressed Marshall, who said he read it three times, telling Palmer it was the "ablest presentation ever put forth by an American Army or Navy officer." Palmer's tract charged that "Congress adopts uncorrelated fragments of both [army and navy] views, appropriates millions of dollars and calls it a national defense system." Palmer developed what he termed a scientific formula to unite these fragments into a balanced, unified national defense organization.[10]

The search for an equilibrium between naval and military means dominated Marshall's activities during the late 1930s. He worked with Assistant Secretary of War Louis Johnson to expand the U.S. Army and, as he told an American Legion Convention, to create a "symmetrical modernization and development of our military National Defense system." As army chief of staff in 1939 Marshall tried to educate the president about this "symmetrical" program. He ordered the commandant of the Fort Benning Infantry School to show FDR something that "would awaken his interest in the practical training of the Army, just as the Navy has stimulated his interest over their big annual maneuvers." Only a strong executive, committed to a balanced defense system, Marshall recognized, could protect army interests.[11]

Marshall reactivated the long-retired Palmer in 1941 as a "military elder statesman" and personal advisor. Though seventy-two years old, Palmer contributed with great vitality to the organizational ferment that accompanied Marshall's service as chief of staff. He resumed earlier ties with Wadsworth, his ally in drafting the National Defense Act of 1920, as the New York congressman legislated the Selective Service Act and began to consider postwar military organization. Palmer cultivated new allies in a group of civilian consultants, social science scholars, and reserve officers who entered the mobilization bureaucracy in 1941–42. These included W. S. Culbertson, E. M. Earle, Harold Sprout, Nicholas Spykman, Father Edmund A. Walsh, and E. Pendleton Herring. "We are working on slightly different aspects of the same problem," Palmer wrote Earle, but he urged collaboration between the military and civilian mobilizers. Palmer explained that they had to correct the defects in the separate system of War and Navy departments and that they had to do so now, to keep from plunging the military back into its segmented and ineffective peacetime establishment.[12]

Marshall assisted Palmer by developing the Special Planning Division of the War Department Special Staff Agency. This division became the army's primary agency for reconversion, demobilization, and single-department planning under Brig. Gen. William F. Tompkins, Col. Gordon E. Textor, and nearly fifty officers. The division reflected Palmer's advice more than any other, probably because it represented Marshall's views. In his studies for the Special Planning Division, Palmer stressed the "one single architectural plan" for national defense as the basis for any postwar planning. He circulated this concept among top War Department officers, including ASF chief Brehon Somervell and George Grunert, deputy chief of staff for service commands. Somervell reflected the growing internal War Department struggles between ground and air officers when he responded that Palmer's single-department design might force the army air force to "adopt a more businesslike and reasonable attitude than it has in the past." Significantly, Somervell expected the Navy Department to cooperate in the single department movement. "The views which I have been able to collect in the Navy," he explained, "are very much to the effect that a Department of National Defense would be a desirable thing."[13]

Palmer's advice influenced Tompkins, who accepted the single-department concept as basic army reconversion doctrine. "Demobilization planning cannot be properly directionized until we have resolved the broad outline of the future permanent military establishment," Tompkins decided. That outline included amalgamation of four "co-planular" groups for air, land, sea, and supply services into a single agency under one executive. Tompkins developed such a blueprint for Marshall in mid-August, 1943, including in it Palmer's "single architectural plan" with a permanent department of defense, chief of staff, and unified supply service. "The benefits of a single department in obtaining unity of command, and elimination of duplication and overlapping, and economy in supply and manufacture are being so clearly illustrated during the present war period," Tompkins wrote Marshall, "that it is, perhaps, the time to determine and formalize the War Department attitude on this question."[14]

Marshall hesitated. He discussed the single-department idea with Roosevelt during the Quebec Conference, but the president opposed such planning while the war was still on. Marshall warned his staff to keep single-department activity secret, and his deputy, Joseph T. McNarney, thought that this secrecy should extend to the secretary of war and White House. Palmer also worried that the navy would learn about what they were doing. He wrote Professor Earle that this secret plan would create "a worse jumble" than that which occurred after the first world war. "Such a proposal would probably be opposed by the Navy," Palmer wrote, "and in any event, would arouse much ill-informed popular controversy that might obscure the real question at issue—that is, the importance of establishing a better, more economical coordination between the several elements of the national defense."[15]

The Navy Department knew about the plan. "Two or three members of my own staff told me that by chance they had seen a large diagram showing what was to be worked out," Adm. Ernest King recalled. That diagram displayed how far War Department thinking had developed toward using the war crisis and reconversion period to institute a single department of defense. One day after Tompkins's memorandum advising Marshall to go ahead with the single-department plan reached the chief of staff, Navy secretary Frank Knox complained to the president about supposed War Department intrigue to legislate such an organization before the war ended. Knox asked Roosevelt to stop this activity.[16]

The navy gave some thought as to how it would approach reconversion but did so in a more piecemeal fashion than army planners did. During the summer of 1943, in response to unrest among naval aviators, King asked Adm. Harry E. Yarnell, who like Palmer had been called out of retirement to active duty, to head a special postwar planning section in the CNO's office. King ordered Yarnell, long ago the first commanding officer of the *Saratoga,* the navy's original full-fledged carrier, to study naval aviation policy and organization. Occurring within moments of the navy's discovery of the army unification diagram, Yarnell's assignment held wider implications than simply a study of naval aviation. In fact, Yarnell surveyed overall national defense organization, and his questionnaire to hundreds of senior naval officers solicited opinions about not only aviation but also the single-department issue. "It is desired to study this problem primarily from the point of view of national defense as a whole," he told Adm. William F. Halsey, Jr.[17]

For years Yarnell had favored one central defense system, since even before World War I he had written articles proposing centralized organization and planning. He had served on the staff of Austin Knight, president of the Naval War College, in 1915 when Knight, a naval reformer and insurgent officer, had called for the merging of the War and Navy departments into a unified structure. At the time, Knight had introduced a concept that later dominated World War II reconversion politics. "With complete cooperation here, a spirit of cooperation which, however, must be made concrete by definite regulations," Knight had announced in 1915, "the question of national defense will at once take on a character of *unification* which at present is totally lacking." Yarnell solidified his unification thinking while on the staff of Adm. William S. Sims's London Naval Planning Section during World War I. Another insurgent officer, Sims advocated a single department as well. When Yarnell returned to Washington in 1919, he had joined Palmer in the reorganization of the Joint Army Navy Board. Like Palmer, Yarnell became intimate with Earle and other national security scholars. In fact, no other naval officer came closer to Palmer on the single-department issue, as evidenced by Yarnell's article endorsing a single department of war, published in August, 1943, in the Naval Institute *Proceedings*.[18]

Yarnell used his special assignment to advance postwar national defense reorganization and unification. He never received an order from King to examine the single-department issue but apparently took it upon himself to ask in a survey, sent to over three hundred senior naval officers, about their views on a postwar single department. One hundred twenty-seven officers responded, including those in the middle of combat operations. All assailed the AAF for blatant wartime publicity that promoted air power, a separate air force, and anti-navy feeling among the public by portraying the sea service as a reactionary branch that delayed winning the war by preventing the full application of air power against the enemy. Many responses favored some form of unification as the only method to curb the growth of an independent air force. Moreover, the war had brought sea, land, and air forces together as never before, accentuating the urgent requirement of coordination in planning and joint operations. "To obtain such coordination there must be unified command, unified indoctrination, and unified training of all the elements representing land and sea power," Adm. David LeBreton told Yarnell.[19]

Encouraged by what he saw as growing support for the single department, Yarnell circulated his recently published article. He asked Professor Earle to distribute copies to army officers, including Col. W. S. Culbertson and Col. Malcolm W. Moss, chief of operational intelligence for AAFHQ. Both expressed enthusiasm. Moss thought the article path-breaking: "I think it is extremely interesting and encouraging that a man of Admiral Yarnell's standing in the military profession should advocate the measures which he brings out in the enclosed article." Naturally, Palmer liked Yarnell's article. He told the admiral that he agreed with every word. Palmer recalled the old days when, in 1919, the two enthusiastic, activist officers had returned from Europe as "co-conspirators" to reorganize the Joint Board "as the first step in the long evolution toward a coordinated national defense system."[20]

Palmer assumed incorrectly that Yarnell served Admiral King as an intimate, influential senior advisor in the same fashion that Palmer operated as Marshall's "military elder statesman." He introduced Yarnell's article to Tompkins as though it contained the official Navy Department position on a single department of war. He recommended that they hold a conference with navy planners to discuss the Yarnell plan. "I am delighted that a concrete plan comes from the Navy," Palmer wrote Gen. Sherman Miles. "In our preliminary studies," wrote Palmer, "Tompkins and I have feared that the Navy would oppose it."[21]

Fortified by Yarnell's article, Tompkins and Palmer persuaded Marshall to introduce his single-department plan to the JCS. Tompkins told the chief of staff on October 11 that the moment had arrived to send the memo forward. The last barrier had been removed. "It is known that at least many ranking Naval officers agree," he assured Marshall. "For example, Vice Admiral Robinson recently informally suggested a single department in the postwar period, and Admiral Yarnell, in charge of postwar planning for the Navy, advocated a single department in an article published in the August 1943 issue of the *U.S. Naval Institute Proceedings*," Tompkins wrote. After several revisions to modify but not delete language on a general staff, chief of staff, and unified procurement service, Marshall approved the single-department plan. On November 2, 1943, he submitted to the secretariat of the JCS his proposal for a single department of war.[22]

There was nothing precipitous or conspiratorial about Marshall's approach to the single department. He had thought about it for years, testified for it before Congress, and discussed it frequently with Palmer. His views on the subject were well known. Marshall did not believe his memorandum posed a threat to his carefully cultivated interservice cooperation because he thought that the debate would remain secret. Nevertheless, Marshall's introduction of the single-department proposal in the middle of the war seemed premature. Probably without Palmer's activity, including his reunion with Yarnell, Marshall would not have rushed the plan to the JCS in 1943, particularly since the navy reacted violently to the memorandum.

Admiral King sent unsigned copies of Marshall's single-department memorandum to top naval officials, explaining that the enclosed plan was of such vital importance to the naval service that he wanted a prompt and formal evaluation. The immediate opposition arose from the navy's traditions, strategic role and mission, budgetary requirements, and organizational perspective. In terms of the larger debate then emerging over reconversion, the navy's opposition reflected an outlook stemming from its decentralized system of bureaus, boards, and administrative forms. In fact, naval organization seemingly represented the coordinate system favored by corporate organizers such as Baruch and Eberstadt. The navy thus offered, in a way, a corporate model, from its functional units integrated and coordinated by a central office at naval operations, to its Baruch-like corporatist body of naval elder statesmen embodied in the General Board.[23]

One of those senior advisors on the General Board, Adm. Gilbert J. Rowcliff, responded first to the single-department memorandum. Rowcliff declared that the unsigned proposal shown him by King revealed an "un-American" plan by the army air force, New Deal agitators, and government "budgeting authorities seeking ways to circumscribe expenditures." Greatly upset by what he saw as a plot to replace sea power by air power through forced centralization, this battleship admiral—once President Taft's naval aide—rejected the entire premise of the single department. Rowcliff called the plan incompatible with the American military experience of separate organizations, traditions, roles, and missions. Adm. Thomas Hart, chairman of the General Board, agreed and urged further study on the entire question of postwar reorganization before anyone considered such a drastic step as merging the two services into one executive agency. In the end, JCS chairman William Leahy demanded postponement of the issue until after the war. "I am afraid that this means that Admiral Leahy wishes to throw the whole question of a single organization back into the study stage," Tompkins lamented.[24]

After hearing these and other objections, King recommended that the Joint Chiefs consider not only the single-department plan but also plans for two and three military departments. He insisted upon this condition as his price for cooperation in the formation of the special JCS Committee on Postwar Reorganization. King's attempt to waylay the single-department movement angered Marshall. "In my opinion," he told King, "no useful purpose will be served by further committee study of the broad question of whether we are to have one, two or three defense departments." Further study wasted time, created interservice friction, and weakened American defense. For the sake of harmony and the prospect of continued consideration of a single department plan, though, Marshall acquiesced in King's demand as "a satisfactory basis for initiating the study." This agreement led to the appointment of the JCS Special Committee on Defense Reorganization under Adm. J. O. Richardson that in the spring of 1944 began months of interviews with top military and naval officers in Washington and in the overseas theaters on the question of the single department and postwar reorganization.[25]

Already Marshall's memorandum had stirred interservice differences. Still, it might have had no more impact than any other memorandum or proposed plan in the politics of reconversion had not Palmer leaked news of the single-department planning to Congressman James Wadsworth, "one old time co-conspirator to another." Palmer urged his friend to organize congressional hearings on the single department, UMT, and related postwar defense matters. Wadsworth welcomed the advice and was determined that Congress reassume its share of leadership over the composition and management of military forces, which had been lost to the executive during the past decade. Wadsworth arranged for the formation of a select committee to investigate postwar national defense organization under the chairmanship of Virginia congressman Clifton Woodrum.[26]

In assembling the agenda for the Woodrum Committee inquiry, Wadsworth dismissed at once the topic of air force autonomy as too controversial. Then, he considered universal military training. Palmer thought that the committee should postpone hearings about this topic until they had had an opportunity to mobilize "broader opinion to offset the well-organized and highly vocal special interests" that included the National Guard, ROTC, and pacifists. Wadsworth, whose own national service bill remained deadlocked in committee, removed UMT from the agenda. That left the single-department topic. As the framework for the reconversion of national defense to peacetime the single department made sense to Wadsworth. The questions of postwar military policies, the size of peacetime forces, the rate of demobilization, the formation of an independent air force, and congressional-executive relations over defense all revolved around the fundamental issue of whether or not the American defense establishment resided in one great central department or in the traditional separate departments. The single-department idea stood close to the ideal of defense organization held by Wadsworth's friends in the War Department, who were committed to this framework for reconversion planning.[27]

The single department proved nearly as controversial a topic for the agenda as universal training or an independent air force. Jolted by the attitudes of Admirals King and Leahy, Marshall believed that beginning the hearings with the single-department issue would provoke public conflict that would prejudice hard-won interservice cooperation. Secretary of War Stimson claimed that the War Department had not yet "incubated" a single-department plan, while his assistant, Robert Lovett, thought that those in the inner circles of the defense bureaucracy ought to settle the question of postwar reorganization among themselves and without congressional interference. All worried during late April, 1944, that Secretary Frank Knox, following the lead of his admirals, might object to the hearings on a single department and refuse Navy Department cooperation.[28]

Knox had grown tired of the incessant interservice squabbles and duplications. He had initiated his own investigation of defense reorganization, directing Raleigh Warner, a Chicago oil executive and organizational expert, to report on management problems. In December, 1943, Warner recommended the combination of all intelligence offices and other joint functions. Moreover, Knox had started to place his department within a larger national defense setting, explaining to one pro-navy congressman that the navy needed to adjust to overall military requirements. "If ever there was a time when the nation needs a minimum of partisanship it is now, along with the maximum amount of unity," he explained. In meetings with Stimson and Woodrum Committee organizers, Knox expressed a willingness to cooperate with the War Department in the investigation of postwar reorganization and the single-department matter. Knox "commented on the inevitability of a single Department of War after the war," his assistant, John H. Dillon, recalled.[29]

Then as the Woodrum Committee prepared for its first session on April 24, Knox became seriously ill. Forrestal's public relations man, Eugene Duffield, warned that the undersecretary might have to substitute for Knox and present the navy's position before the committee. Forrestal had expected to follow Knox's opening statement with a brief explanation of naval procurement, stressing the importance of its separation from army supply. Now he would lead the testimony. Forrestal was not worried. He told Eberstadt that the hearings would be "good and fair" because Carl Vinson, Congressman Walter G. ("Ham") Andrews, and other staunch navy friends sat on the Woodrum Committee.[30]

Despite Forrestal's sanguine observation, the inquiry on postwar defense deteriorated into a forum where the army lobbied aggressively for a single department with one chief of staff and one defense secretary. It was as if Marshall's long pent-up frustration with the two-department system had exploded at last. Testimony reflected Palmer's dominant influence, from the specious idea that the single-department plan originated with George Washington's sentiments on a peacetime military establishment to the explanation that unity of command and centralization were in the American military tradition. Army spokesmen insisted that the War Department stood behind the single department unanimously. Deputy Chief of Staff Joseph McNarney concluded the army's presentation forcefully, using those elaborate organization charts that King's people had spied earlier pleading for "one great unification."[31]

McNarney's "one great unification" slogan caught on with press and public, focusing attention disproportionately on this aspect of reconversion. Unification implied much more than combination of army and navy into a single department. In the public forum, unification became the term that encompassed the entire struggle over postwar national defense reorganization, the growth of executive power, and emergence of a national security policy and structure. No longer a matter of secret JCS committee discussion, the topic generated for the nation at large images of civilian-military conflict and of a Congress-versus-president power struggle over reconversion policies. It focused on a fight between advocates of sea power and of air power for postwar defense leadership and strategy. Most important, the unification issue, kept alive in the press, continued the battle over postwar organization and political-economic policy between New Dealers and dollar-a-year types that had started during prewar mobilization, grown during the wartime production debate, and accelerated during the early phase of reconversion planning.[32]

Knox's illness and the emerging public debate over the "one great unification" pulled Forrestal directly and abruptly into the politics of reconversion. Almost immediately the pro-unification press cried: "Forrestal is Chief Bar to Merger of Services." Testifying before the committee, Forrestal insisted that the navy "definitely" opposed legislation for a single department of war. He called the movement for reorganization premature, just as he had considered early reconversion to civilian production premature. Reflecting General Board advice, he advocated further study on what he termed the radical proposal. Forecasting future problems, Forrestal demanded that in any postwar reorganization the navy must retain its own aviation arm and the marines.[33]

The undersecretary went beyond General Board advice (and probably beyond Knox's view as well) when he introduced to the proceedings a corporatist tone that promised to make the anti–single-department position part of a larger debate against those who favored more governmental controls and regulations. Forrestal explained to the committee that modern corporations had adopted a decentralized organization much like that of the Navy Department and two-department system. They knew that in the private business world consolidation under a single executive rarely fostered better efficiency or greater savings. In an ironic forecast for the man who in 1947 became the beleaguered first U.S. secretary of national defense, Forrestal testified in 1944 that the current defense organization was far too complex and diverse for any single executive to administer and manage. "There is no human being capable, in my judgment, of sitting on top of all of that and assuring that you have the fine integration and efficiency which it is presumed would result from consolidation," Forrestal argued. He urged instead that the committee consider a corporate system of defense that had appeared in 1925 in the Morrow Board Report. The Morrow Board, comprising national statesmen and managerial experts, had recognized the unique missions of land and sea aviation and other activities and of separate departments for war and navy. The Morrow experts had stressed the importance of collaboration between industry and the military and between business and government through formal and informal organizations. They had presented what might be seen as a corporatist strategy for national defense organization.[34]

Knox died two days after Forrestal's appearance before the Woodrum Committee. Forrestal seemed the natural successor. However, the undersecretary expressed reservations about the top job. Apparently, he preferred the industrial mobilization post to the more political duties of a department secretary. Corcoran urged Eberstadt to ask their friend to stop saying that he did not want to be navy secretary or the president might believe him. Meanwhile, Harold Smith, budget director, who at one time recommended Forrestal for economic director, now opposed his elevation to this top wartime Cabinet post. Smith's opposition reflected the hardening lines between New Dealer and corporate organizer, with Forrestal representing the latter through his close association with Eberstadt, Baruch, and dollar-a-year mobilization experts. Forrestal's recent opposition to the single-department format at the congressional hearings placed him in direct conflict with Smith, whose defense of the single department before the same Woodrum hearings was called by army observers the most brilliant exposition of the entire session. Smith and his budget experts had become firm allies of Marshall and War Department planners in their internal reorganization and in planning for army-navy consolidation. Consequently, Smith urged FDR to bypass Forrestal and to appoint Wendell Willkie or someone with more political visibility to fill Knox's position.[35]

Roosevelt settled on Forrestal, who took the oath as navy secretary on May 19, 1944. Forrestal's experience qualified him better than anyone to continue the job of mobilizing the navy for its final drive across the Pacific. Forrestal's promotion ushered in another period of intense activity as he refined departmental management structures for industrial mobilization and civilian-naval collaboration. He introduced innovations such as the Top Policy Group and research and development agencies. He expanded a public and congressional relations program designed to publicize the navy's role in winning the war (and not incidentally in defending the peace after the war).[36]

The public relations aspect of Forrestal's job took an inordinate amount of precious time that needed to be spent on his primary task of mobilization and war. Reconversion politics intruded constantly. He explained to Senator Truman, a leading proponent of Nelson's plan for early reconversion to civilian production in order to protect and expand small business and labor, that such premature reconversion endangered military production, destroyed national morale, and delayed victory. Forrestal wanted them to concentrate on the crisis at hand. But the navy secretary found Truman unsympathetic to his plea for a delay to all reconversion planning and programs. Forrestal admitted that he "had not made any great headway" with Truman.[37]

Truman figured as well in Forrestal's growing involvement with the unification phase of reconversion. In 1944 Truman became the most outspoken proponent of the single department of war, which he discussed in "long talks with Marshall." Truman's own experience as an artillery officer during World War I, as a member of the Missouri National Guard, and as current chairman of a Senate committee that had saved the taxpayers millions by eliminating military waste and duplication had given him a perspective on the two-department system very similar to that developed by Marshall. Truman found the traditional separation of the services inefficient and expensive. "A stiff-necked contentiousness still marks Army and Navy contacts in the lower echelons and will continue to mark them as long as each arm enjoys an independent status," Truman observed. He announced proudly that he had written the American Legion resolution in favor of a single department and supported publicly one general staff "in full charge of tactics and strategy viewing the nation's offense and defense as an indivisible whole, and totally unconcerned with service rivalries." He also wished to eliminate separate officer training programs and entrenched traditions that he felt bred elitism and military cliques. "This seems to be a fixation with him," Forrestal observed.[38]

Shortly after Roosevelt selected Truman as his vice-presidential running mate in July, 1944, the senator published an article, "Our Armed Forces Must Be Unified," in *Collier's* magazine. In this article, Truman traced a long history of bitter interservice rivalry that he felt had hindered operations during the Spanish-American War and World War I. "In the last war," he wrote, "it was a great joke that the Navy got the jump on the Army, filling warehouses beyond all possible need and hoarding in spite of the Army's pressing necessities." Truman also blamed the Pearl Harbor disaster on the separation of army and navy, with its division of authority and two sets of orders. Drawing on long-held views of public administration and presidential control, Truman offered centralization as the solution to overcome duplication, waste, and inefficiency. "The end, of course, must be the integration of every element of America's defense in one department under one authoritative, responsible head," he concluded. "Call it the War Department or the Department of National Security or what you will, just so it *is* one department."[39]

Truman's *Collier's* article disturbed Forrestal. Four days after it appeared in the popular magazine, Forrestal wrote Congressman Vinson, "I do not think for a moment we can take [the question of a single Department of Defense] lightly, and I have so told Admiral King." The secretary rushed an assistant over to a White House press conference to learn whether by selecting Truman FDR had changed his position on unification. Forrestal wanted the "exact language" that Roosevelt used when responding to that question. The secretary began to mobilize the naval establishment in September, 1944, to fight the single-department issue. He explained his decision to Adm. Chester W. Nimitz, commander-in-chief of the Pacific fleet. "The question of a single department of national defense—or rather, the form our national military organization is to take after the war, will clearly be up for vigorous debate during the next eighteen months," he wrote

Nimitz. "Whether we like it or not, there is no question but that the Army through its great size and its multiple avenues of communication and the aggressive publicity activities of the Air Forces has had a profound effect on public opinion," he wrote, then asking Nimitz to develop the navy story of the war. "I would not impose this task upon you except for my compelling conviction that the future existence of the Navy as an independent military organization depends upon the public's appreciation of the Navy during the next several months."[40]

Securing public appreciation of the navy was not enough. Forrestal sought to convince Congress and the president to enact separate appropriations for the postwar navy before the single-department movement gained popularity as the solution for postwar defense. He had to present "a concrete statement of the Navy's view," Forrestal told Massachusetts congressman Christian Herter. So far, the Navy Department had not developed any concrete program for reconversion and peacetime. He worried to King about the "inchoate condition of our postwar studies," instructing the admiral to assign a special group of officers to examine naval base requirements, fleet strength, reserve labor, and relations with industry in the preparation of a "concrete statement of the Navy's views." Forrestal was particularly anxious to have a document ready for the president because he felt that as long as FDR remained in the White House, the navy would be protected.[41]

Forrestal wrote Roosevelt on April 9, 1945, that he wanted to send several officers to Warm Springs, where the president rested, to make a twenty-five minute presentation complete with tables and charts that explained the navy's postwar needs. "It is only a framework against which to plan," Forrestal admitted, "but I think it is sufficiently tangible to be interesting to you." The navy representatives never made the trip to Warm Springs, where Roosevelt had died three days after Forrestal had dispatched his note. Like Knox's death the previous year, FDR's passing accelerated the pace of Forrestal's involvement in the unification controversy. It brought to the White House the man who had crusaded publicly for the army's single-department plan. Though Forrestal had endorsed Truman for vice-president "in view of the fact that the alternative would be Henry Wallace," his views differed greatly from Truman's on reconversion and unification. Forrestal considered the new president a threat to the navy's interests (and probably to those of big business as well). Fearing that Truman contemplated immediate armed services consolidation through an executive order, Forrestal and Admiral King visited the White House a few days after Roosevelt's death to present Truman with the navy's views on postwar organization. At the end of the meeting, Forrestal urged the former Missouri senator to "reread the Morrow Board Report with the thought that this form might be followed in the study of the desirability of consolidating the two Services."

Two days later, Forrestal sent his personal copy of the Morrow Report to the president, "on the chance that you might not have made a mental note of your decision to look at it again."[42]

Truman was not interested in the Morrow Report. Not only had the Morrow Commission recommended in 1925 against a single department and independent air force, two forms that Truman warmly endorsed, but it had contained those outside corporate organizers that during the war Truman called "hostages to fortune" and under "the controlling officials of their corporations." The Morrow Board of experts seemingly represented those elitist "cliques" that Forrestal found had become a fixation for Truman. The president wanted all policy-making and administrative functions to reside in regular governmental channels and be carried out or supervised by elected representatives. "If proper progress is to be made, it is necessary to permit the President to lay out the machinery for carrying out his responsibility for the conduct of the executive branch," Truman argued. Executive responsibility included the determination of defense organization and policy. The Morrow Board recommendations implied a coordinate form of government by and through an array of public and private, formal and informal committees, joint commissions, and "extracurricular" agencies. Truman wanted such forms absorbed into regular cabinet government and their coordinating functions assumed by the federal bureaucracy. Undoubtedly Truman agreed with Nelson's observation during the war that such committees "get together to see how they can screw the public." Truman's views on organization seemed destined to run head-on into the war-enabled corporatist organizational schemes.[43]

Forrestal, on the other hand, endorsed the Morrow Board's report as the type of expert advice that enabled government to develop orderly policies without the disruptive influence of political conflict and individual interests. It recommended structures "similar to the British system for coordinated and focused government action," Forrestal explained to Harry Hopkins. Instead of the regular departments absorbing and burying the "extracurricular" forms, as recommended by Truman, Forrestal wanted these emergency wartime agencies to become permanent. He advocated a civilian JCS, peacetime WPB, and full-time State-War-Navy Coordinating Committee to tie together and coordinate existing departments and offices of government.[44]

When Forrestal asked Truman to reread the Morrow Report in April, 1945, their differences over philosophy and government organization were neither as distinct nor complete as they later became. Introducing the contrast at this point permits the emerging conflict between Forrestal (with Eberstadt) and Truman over unification to be more clearly understood. In this bureaucratic battle, Truman's philosophy, which had determined his support for strong federal regulation and executive control over mobilization, war production, and reconversion programs, helped to explain his advocacy of a unitary principle of national defense. Likewise, Forrestal's quasi-corporatist inclinations (greatly reinforced and developed by Eberstadt), which had guided his support of the ANMB reorganization, Controlled Materials Plan, and other mechanisms for coordinating business, industry, and government, determined Forrestal's defense of a coordinated system of national security.[45]

Forrestal framed his opposition to department unification in terms of naval interests rather than corporatist organizational principles. He realized that the debate meant a struggle for the "system of coordinated and focused government action" against continued New Deal centralization of state authority that he worried could lead to socialism and dictatorship. The United States had all the ingredients for "the creation of a real socialist state," Forrestal warned Hopkins. One of those ingredients was the relentless consolidation of power in agencies such as the single department of defense. Forrestal's fight against unification suffered a severe setback with FDR's death and the beginning of the Truman presidency. Another setback occurred coincidentally on April 12, when the JCS Special Committee on Reorganization released its final report that advanced everything that Forrestal and the navy opposed, including the single department, chief of staff, and secretary of defense.[46]

The JCS report devastated the Navy Department. Senior naval officers wondered how the Joint Chiefs could release such a disruptive report, just as the navy made final preparations to hit Japan's inner defenses, for the greatest naval and amphibious assault in U.S. military history. The single-department idea had to be suppressed, King's chief of staff advised. It was an instrument for "extra-constitutional" control over the national defense by the army and air force, Adm. Joseph S. McCain warned Forrestal. Admiral Leahy thought the report radical and dangerous, while the acting chairman of the General Board, E. E. Kalbfus, called release of the report a grave error. Kalbfus advised Forrestal to stop the recommendations outlined in the report by "all legitimate means." If this report received President Truman's endorsement, Adm. Samuel M. Robinson warned Forrestal, "there is no telling what this might lead to."[47]

Navy Department reaction to the JCS report—and indirectly to the fact that a former army artillery captain now sat in the White House—peaked at the Navy's Top Policy Group meeting of May 7, 1945. Here, in an emotional and angry debate, civilian and naval leaders attacked what they saw an an army plot to sneak unification through during a period of extreme national crisis and total war. Hensel wanted to publicly condemn the War Department for its single-department plan and tactics because, as he claimed, "the Army has been pursuing a very definite policy of slipping stuff out the back door." Capt. George L. Russell, in charge of postwar organizational planning in King's office, compared the JCS report to the transcript of a court-martial with the navy on trial for its life and found guilty before the evidence had been presented. "Before they get to putting us under the Army, they better find out what the results will be," said Frederick Horne, Vice-Chief of Naval Operations. "I think we have handled the Army too lightly in view of what they have said we have or haven't done," Horne added. He wanted the navy to take off its kid gloves and fight back.[48]

How could the navy fight back? Forrestal's legal assistant, Mathias Correa, urged the secretary to convene a special committee to prepare arguments against the single department. Hensel recommended that they get Baruch to head such a committee. Instead, Forrestal sought to settle the problem through personal mediation. He approached the secretary of war and asked him to put aside the single-department madness and work together for reorganization. He appealed to Stimson's corporate instincts and legal background. "It seems to me that if we could find an accommodation through which this latter subject could be approached by the two Departments in an effort to find a solution between themselves rather than submitting it to the forum of Congress," he explained to the secretary of war, "we should be making a sounder use of our time." At the same instant, Forrestal invited King, Marshall, and Hopkins to lunch at his house to solve the problem among themselves. He also contacted Carl Vinson, David Walsh (who chaired the Senate Naval Affairs Committee), and other congressional leaders to assist in reaching a compromise. He sought the support of Walter Lippmann, Palmer Hoyt, David Lawrence, and other sympathetic journalists to prepare public opinion for such a compromise.[49]

During the spring of 1945 Forrestal's activities to find a compromise to the single-department issue before the dispute became public projected the image of the secretary as a Washington operator. His vigorous leadership, one admiring observer claimed, set an example for the Navy Department unequaled since that made by Alfred Thayer Mahan, the nineteenth-century philosopher of sea power. Instead of becoming a great operator who guided the bureaucratic politics of reconversion, however, Forrestal stumbled into a unification dilemma. Stimson rebuffed overtures for private accommodation. He told Forrestal that if the War Department did not press on with its single-department plan, then "the golden moment for accomplishing what I consider is a great contribution to the effective and economic defense of the country, will have passed." At the same time, Forrestal's luncheon failed to bring Marshall and King together. The chief of staff remained "unshakably committed to the thesis of a single civilian Secretary with a single military Chief of Staff," Forrestal discovered. Neither Forrestal's congressional allies nor press friends seemed able to stem the movement toward the unified defense establishment.[50]

It became obvious that Forrestal could not reach a settlement through personal intervention. Moreover, he had no alternative plan to help the navy offset the army's well-developed single department scheme. Woodrum told Forrestal during the hearings how disappointed he had been that in opposing the War Department's blueprint, Forrestal had offered no ideas of his own on overall policy and organization.[51]

Partly, Forrestal's lack of a clear alternative stemmed from his tendency to reflect on issues to the point of near paralysis. He "always got involved with all kinds of complicating qualifications and reservations," one wartime assistant observed. Forrestal's assistant, Eugene Duffield, and journalist friend Walter Millis described the secretary as having a sensitive, restless, searching mind. Forrestal mulled over problems, elicited advice from numerous sources, weighed the pros and cons, changed his mind, and delayed making a final decision. Eberstadt understood his friend's intellectual process when he told one journalist that Forrestal was not the "chief bar to merger of the services" because he had not made up his mind on the unification issue. Forrestal needed time to study, reflect, and seek more advice. "The position I have taken, which I am confident was Frank Knox's," he told Walter Lippmann, "is that it should have the most objective and searching examination possible–that the conclusion should be reached after the facts are in rather than producing the facts to implement a conclusion."[52]

The wide-ranging Forrestal inquiry covered everything from the fate of postwar Germany to American control over the Japanese mandated islands, from British cabinet government to dialectic materialism and Marxism. He had reflected constantly upon the nature of postwar economic, defense, and security organizations since arriving in Washington nearly four years earlier. He talked about world issues, domestic politics, and national defense regularly with Eberstadt. He clipped newspaper editorials and articles and recorded quotations from books and conversations in a set of notebooks. He was familiar with scholarship on national security and urged Princeton University to develop a course on the fundamentals of national power and security. He gossiped about postwar international organizations and policies over lunch and dinner with a wide array of guests, including John McCloy, Robert Lovett, J. F. Dulles, Henry Luce, Bernard Baruch, Arthur Krock, Walter Lippmann, Harry Hopkins, and visiting British statesmen such as Anthony Eden and Lord Halifax. He communicated frequently with Ambassador W. A. Harriman in Moscow about the emerging threat, as they saw it, of Soviet-Marxist postwar expansionism.[53]

Forrestal tried to absorb so much and develop such a sophisticated worldview that some accused him of having only superficial understanding of issues. He quoted constantly from the theories of Walter Bagehot, a nineteenth-century English constitutional scholar, Dean Acheson observed, but showed little awareness of how these ideas applied to current problems. Forrestal borrowed bits and pieces of ideas from a variety of sources including the American liberal-corporate business world and British parliamentary government. He adopted the naval aviators' view of sea-air power, creating an awkward Mahanian hybrid approach to the Navy's role in national defense. In his constant reflection, he learned from Progressive Era administrative formulae presented by his friend Harold Dodds and other Wilsonian scholars. The neo-realist concepts of Dulles and Eberstadt, deeply impressed Forrestal, too, including their emphasis on Metternich's role in creating stabilization through a postwar Congress of Vienna system. In the end, those fragments remained unconnected, never systematized by a coherent organizational or political-economic philosophy.[54]

Nevertheless, throughout his life at Princeton, on Wall Street, and as a public official, Forrestal's strength stemmed from an ability to find and use as advisors the kind of managers and experts on the Good Man List, individuals who in moments of crisis sorted out tangles, developed systems, and prepared appropriate courses of action. In May of 1945 the unification phase of reconversion had reached crisis proportions for Forrestal. He resumed his daily telephone calls and personal meetings with Eberstadt, which had been interrupted shortly after his friend's removal from the War Production Board in February, 1943. Though they had communicated periodically during the intervening years, Forrestal and Eberstadt had not maintained the intense contact that between late 1940 and early 1943 had given Forrestal strength and guidance. Now Forrestal's calls to Eberstadt in New York City, and the latter's journeys to Washington, increased once again. In late May, Forrestal discussed the single-department crisis with Eberstadt, Captain Russell, and Douglas Dillon, the son of his former business partner and now a naval reserve officer on duty in the secretary's office. A few days later, Eberstadt occupied an office near Forrestal's and began examining the entire unification problem. Once again Eberstadt prepared to give Forrestal the direction, the system, and the firm resolve that he lacked.[55]

6. A National Security Plan, 1945

When Donald Nelson, WPB chairman, fired Eberstadt in early February, 1943, the latter swore never to return to Washington's bureaucratic battles. Behind his tough exterior lay a sensitivity to criticism and an ego that bruised easily. With unhappy memories of his WPB experience still fresh in the spring of 1945, Eberstadt hesitated to rejoin Forrestal in an already smoldering unification controversy, where the Truman administration carried on the kind of factious politics that had driven him out of Washington in 1943. "You, I think, know very well that I had a bellyful and a mighty sour one at that," he wrote Forrestal.[1]

The navy secretary persisted. Driven to near panic over the single-department movement, Forrestal needed Eberstadt's steadying counsel. He called Eberstadt regularly in New York City during late April and early May, inviting him to Washington for private consultation immediately following the tumultuous May 7 Top Policy Group meeting on unification. From May 15, when David Walsh, chairman of the Senate Naval Affairs Committee, urged Forrestal to present an alternative to the War Department proposal, until Eberstadt began such a study on June 19, Forrestal called and met his best friend with increasing frequency. His appeals for assistance became more urgent: "What I have in mind is a survey of, and advice to me, on the question of unification of military services into a single department of national defense."[2]

Eberstadt delayed. He realized (perhaps better than Forrestal) that the unification project had deep political-economic and organizational implications that overshadowed the simple combination of army and navy into a single department. Unification was "a pretty complicated subject," he told Forrestal, "on which opinions seem to be more plentiful than knowledge." From the beginning, Eberstadt placed unification in a broader setting and linked it to earlier mobilization and reconversion studies such as the ANMB, Controlled Materials, and Baruch-Hancock plans. He told Forrestal that any unification study needed to include discussion of industrial mobilization planning, labor resources, a policy for occupation of captured territories, and larger international problems such as relations with what Eberstadt termed "the Russian question mark."[3]

As they discussed preliminary plans for a unification study, Eberstadt gave Forrestal an article from the conservative British journal, the *Economist,* that predicted future struggles for world power between the forces of democracy and Marxism. The *Economist* essay insisted that the latter had the advantage because it presented a line of economic philosophy. Developing an American economic philosophy was on Eberstadt's mind as he considered the unification question, and he held definite ideas on a corporative political-economic philosophy that countered the Marxist system. However, Eberstadt knew little about the military aspect of unification. He hesitated to take the Forrestal job because he suspected that his friend was drawing him into "a political and not a philosophical situation," and as one observer noted, "he knew from the beginning that unification was going to be the order of the day regardless of the convictions or desires or reasoned arguments of the naval service."[4]

Nevertheless, Eberstadt took office space in the Navy Department near the secretary in order to assess the unification problem. He served as Forrestal's personal counsel and not as a "Navy advocate," he told Capt. George Russell, Admiral King's single-department planner. He worked as a consultant on unification as he had previously on the machine tool and ANMB reorganizations. He promised to "reach a strictly impartial and independent viewpoint" for Forrestal. "What he wanted was my independent slant on the subject and I tried to give it to him rather than unconsciously echoing his own views," Eberstadt explained later to President Truman's naval aide, Robert L. Dennison.[5]

Still, Eberstadt consulted nearly every day over the next four months with Forrestal by telephone, at lunch, and in twenty personal meetings where they discussed the progress of the study, examined drafts of the report, and ironed out conclusions. While everyone else referred to the final product as the Eberstadt Report, Eberstadt called it the Forrestal Report. C. Douglas Dillon, who worked intimately during 1945 with both men and served on Eberstadt's unification study group, insisted that Forrestal knew what he wanted when he brought Eberstadt down to provide a report on unification. "Forrestal's influence on Eberstadt was very great," Dillon observed, "and I have always felt that Eberstadt was guided largely in his Report by Forrestal's wishes."[6]

In early June, 1945, Eberstadt remained unsure about what Forrestal wanted or how far he wished to go in examining the question of unity. Forrestal was in another crisis similar to the ones before his appearance at the Pecora congressional investigating committee in the thirties or during the machine tool jam of 1940 when he had to have Eberstadt's help. After reading Eberstadt's *Economist* clipping, Forrestal worried about the "depth and extent" of the ideological confrontation between Marxism and capitalism. He had no answers to the problem but wanted Eberstadt's assistant, E. F. Willett, to continue his study on dialectical materialism for him. Forrestal believed that a strong navy as the first line of defense was necessary to protect American interests in this forthcoming confrontation, and he worried that the single-department plan threatened to undermine the navy's ability to defend the United States. He recognized that the immediate need was to develop a positive navy program to offset "the unification ideas suggested by the Army." However, development of his own unification commission under Eberstadt defied administration single-department policy, particularly since President Truman told Forrestal on June 13 that he had definite ideas on national security organization that he would soon make public.[7]

Eberstadt plunged ahead with preparation of a case for Forrestal. He examined a single-department file kept by H. Struve Hensel, the navy's assistant secretary. Its contents revealed the complexity of the unification issue, and the type of documents in the file set the basic outline for Eberstadt's investigation. The Hensel file contained articles by Truman and Gen. J. M. Palmer, studies of German and Soviet defense organization, notes on the British Committee on Imperial Defense, the Strauss-Draper Report on Joint Procurement, Eberstadt's own ANMB report, and copies of congressional investigations and debates on national defense organization. Each in its own way introduced the different dimensions of the unification problem. At the same time, Eberstadt held several conferences with Willett; Bertrand Fox, his former WPB colleague and departmental advisor; and Capt. George L. Russell. After a lengthy conversation and with only lukewarm support from Russell, they decided that the navy secretary required a study group stocked with research assistants to prepare a comprehensive report on unification and postwar national security organization.[8]

Eberstadt already saw an opportunity to design corporatist interlinking mechanisms to connect functional groups involved in production, education, defense, and public information. For his report to offer an alternative to the single-department plan, it needed to present a blueprint for permanent structures that would continually connect military, foreign policy, and economic sectors in a unified national security establishment. During their meetings and conversations between May 31 and June 16, Eberstadt persuaded Forrestal to pursue this type of unification study. On June 18 Forrestal informed Secretary of War Stimson that he had given the question of the single department serious thought and had developed the idea of "an agency of the government to function in both war and peace in an overall coordinating and advisory capacity for both civil and military purposes." The next day, Forrestal hired his personal counsel Eberstadt to begin a unification study. He employed the identical format that initiated the Morrow investigation of national defense organization in 1925 by asking Eberstadt to examine three questions. Forrestal wanted to know whether (1) the single department, (2) reorganization of the current establishment, or (3) the creation of new structures would provide the best national security. To this extent, he set the parameters and frame of reference for Eberstadt's study.[9]

Committed to preparing a unification report, Eberstadt recruited experts on the various elements of national security. One study group member, historian Elting E. Morison, believed that Eberstadt went out and found "a lot of able people so that it looked as though his study was the product of careful and objective research." More likely, Eberstadt borrowed a model from his own experiences on Owen Young's commission of reparations experts and from former law partner Joseph Cotton's close friend Dwight Morrow's committee of experts that prepared the Morrow Report for the president in 1925. Eberstadt knew that Forrestal considered the latter an ideal format. Furthermore, Eberstadt always employed research assistants "to do spade work" for legal and investment clients and now applied this principle to government clients as well. Most important, recruitment patterns and the committee-of-experts concept reflected Eberstadt's corporatist education. He favored formal and informal arrangements whereby "skilled men move back and forth between Washington and their own principal employment." This corporatist concept had as much impact on the character and final outcome of the unification report as the original impetus of providing Forrestal and the navy with an alternative to the single-department plan.[10]

A pool of talented corporate organizational experts, now wartime naval reservists on active duty in Washington, presented a rich recruiting ground. "Eberstadt came to the Department and found out very fast from the Secretary's Office what people might help in the report," Morison observed, "and insisted on getting them from the Secretary's Office, COMINCH, CNO, and the Bureaus." The type of conflict that punctuated Eberstadt's first venture in government service reappeared at once. Recruiting brought Eberstadt into the middle of an ongoing bureaucratic battle over internal Navy Department reorganization that had arisen between Admiral King and the secretary's office during the war. Forrestal warned Eberstadt that opposition to this study would arise because of a power struggle among bureaus and offices for authority. Forrestal said that they should try to get assistants "from the bureaus that have no axe to grind."[11]

Morison claimed that Eberstadt's attempts to recruit experts "caused some ill feeling which existed between civilians and service personnel." Difficulties for Eberstadt arose less from a civilian-military schism than from Admiral King's suspicion that Eberstadt's presence and assignment to prepare a report might adversely influence the postwar role and organization of the navy. Despite Forrestal's assurance that Eberstadt was "a member of the Navy team," King tried to get Eberstadt out of Washington at this critical moment on a tour of Navy facilities and bases, what Eberstadt called a "Pacific junket." Unsuccessful in this direction, King assigned Captain Russell as a liaison with Eberstadt's study group, possibly to watch the civilian and most certainly to indoctrinate Eberstadt to the Commander-in-Chiefs view of the unification problem. "The basic cause of the present agitation," Russell informed Eberstadt, "is the growth of the Air Force."[12]

Meanwhile, Frederick Horne, vice-chief of naval operations, and several other naval officers placed obstacles in Eberstadt's path. They refused to assign to his study group any naval reserve officers who were Eberstadt's personal friends. Thus, they blocked banker Arthur L. Wadsworth's attachment to the Eberstadt group and did the same to Hulbert Tripp, a Wall Street broker and the University of Rochester treasurer, and Richard P. Swope. Eberstadt was particularly upset by navy refusal to release Swope, since he was the son of old friend and Long Island neighbor H. B. Swope. The senior Swope wrote that he was disgusted with the navy bureaucracy for not allowing his son to work with Eberstadt. All these obstacles formed part of a pattern, Willett concluded: "There is some resentment against bringing in outsiders like Mr. Eberstadt to make a report of this nature and importance, and that people with this attitude would like nothing better than to see the whole thing amount to nothing."[13]

Forrestal issued an order giving Eberstadt the freedom to speak with anyone in the naval establishment without prior clearance or going through the chain of command. He helped pry some thirty naval reserve officers from other duties to aid the unification study, including a group that fit Eberstadt's Good Man qualifications and who appeared later on his postwar Good Man List. Composition of this group determined the final product. Only two prominent members of the group were "outsiders": Saint Louis businessman Sidney Souers and Chicago investment analyst Scott Harrod. The rest resided in New York or Boston, had graduated from Harvard or Columbia, and before the war had worked for firms that were associated with the great Harriman and Rockefeller financial interests. They were all the same age, educated formally and informally during the 1930s by the Depression and the rise of totalitarian regimes. Though they rarely met as a single committee, their intimate ties led to frequent informal consultation and discussion. Harvard graduates and faculty dominated the group: C. Douglas Dillon, Maxwell Rabb (assistant to Sen. Henry Cabot Lodge and later a Cabinet secretary during the Eisenhower administration), history professor Myron Gilmore, law professor Milton Katz, and Eugene Homer Catron of Dillon, Read.[14]

Catron was typical of the Good Men on the committee. A product of the Harvard Graduate School of Business Administration and Yale law school, Catron entered the Wall Street law firm of Hines, Rearick, Dorr and Hammond and shortly before the war joined Forrestal at Dillon, Read. After a period in the WPB's Steel Division, Catron became a special assistant for scheduling in the navy's Office of Procurement and Material and an authority on requirements in the Office of the Assistant Secretary of the Navy. The latter position dealt with Eberstadt's Controlled Materials Plan and provided perfect training for work with Eberstadt. Catron wrote sections on logistics and procurement for Eberstadt's report and helped translate Eberstadt's final report into the "Navy Plan for National Security." Under Forrestal's instruction, Catron continued to report confidential departmental business after Eberstadt left Washington, keeping him informed about the work of the secret Secretary's Committee of Research on Reorganization (SCOROR) and on preparation of navy witnesses for testimony before various congressional hearings on unification. No other junior officer worked more actively toward preparing the navy's case against the single department or in promoting Eberstadt's work as the blueprint for postwar national security organization than did Catron. In recognition of this contribution, Eberstadt brought Catron into his investment firm after the war.

Catron assisted in the recruitment of two others who held the organizational expertise that characterized Eberstadt's Good Men. He recommended Dr. Edward Barker, head of the engineering school at Columbia University and a special advisor on personnel matters in the navy secretary's office, to prepare for the report a section on education and training. Catron secured attachment of his "personal friend" Milton Katz. He told Eberstadt that Katz had been "in effect, personal attorney to Mr. Donald Nelson" on the WPB. Eberstadt knew this. He identified Katz with the New Deal element on the board that he believed responsible for his own firing in February, 1943. However, Eberstadt recognized Katz's expertise on the legal aspects of production control and mobilization. From all appearances Katz was a Good Man, and Eberstadt borrowed him from old friend William J. Donovan's Office of Strategic Services. Over the next two months, Katz reviewed the legal issues accompanying the individual organizational recommendations of group members, contributed a section on mobilization, and rewrote parts of the section on intelligence.[15]

Although his recruits determined the shape and texture of the report through their own research projects, Eberstadt recognized from the beginning that he needed an overall scheme, a conceptual framework for the report. He turned for advice to naval reservist Robert Howe Connery, an expert on the history of naval mobilization who was then working on historical projects in the department under Princeton professor Robert G. Albion. From Connery Eberstadt got the idea to broaden the unification study to include all aspects of industrial and resource mobilization. Connery, who received a doctorate in political science from Columbia University and was an associate professor at Catholic University, had served as a research assistant with the Brownlow Commission on executive reorganization. His professional experiences had frequently brought him in contact with those social scientists who moved back and forth between government service, the university, and associations such as the Public Administration Clearing House (PACH), the Brookings Institution, and the Social Science Research Council (SSRC). Connection to the SSRC that published his doctoral dissertation on the National Recovery Administration familiarized Connery with its study entitled *Civil-Military Relations: Bibliographical Notes on Administrative Problems of Civilian Mobilization,* edited by E. Pendleton Herring, SSRC vice-chairman and head of Harvard's Graduate School of Public Administration. Connery introduced Herring's work to Eberstadt.

Eberstadt wanted Herring for the unification study. After negotiating the point of whether the Harvard scholar would do more than acquire volumes for the library–something Herring wished to leave to one of his younger associates, government professor Louis Hartz–Eberstadt recruited Herring in late June, 1945. Over the summer, the two men shared ideas nearly daily on national security concepts and structures. No one saw more of Eberstadt than Herring did. The political scientist gave the report a methodological and conceptual framework. He translated current social science and public administration theory to the receptive Eberstadt. He contributed his extensive research on reorganization, regulatory agencies, interest groups, public opinion, lobbying, and congressional politics. Herring linked their work to that of Woodrow Wilson's generation through the writings of Herring's dissertation director at Johns Hopkins University, W. F. Willoughby, one of the Progressive Era's pioneers on public administration. At the same time, though, Herring reflected the growing disillusionment with Wilsonian idealism among fellow social scientists such as Harvard colleague William Yandell Elliott and the German-trained political scientist Carl J. Friedrich, who ran a seminar with Herring. These intellectual impulses coincided with Eberstadt's own emerging "realist" concepts and long-held corporatist organizational assumptions.[16]

Herring's most significant contribution to the ideas and forms of national security that emerged from the Eberstadt Unification Report of 1945 lay in the corporatist scheme that he developed in his 1941 book, *The Impact of War: Our American Democracy under Arms,* widely regarded as the single most important synthesis of civil-military relations written before World War II. In it, Herring combined traditional military history and studies on sea power, air power, and universal military training with the results of recent social science scholarship. The latter included Harold D. Lasswell's concept of the Garrison State, Hans Speier's work on class structure and total war, Harold Laski's study of the impact of warfare on the modern state, and SSRC chairman Lindsay Rogers's analysis of civilian control over military policy. Survey of this literature had Herring convinced that the greatest problem confronting American democracy was how to mobilize for total war without crushing individual freedom and enterprise with too many controls and regulations. In other words, he envisaged a system that steered a course between fascism and communism and managed labor and material without welding them into the "seamless whole" of totalitarianism.[17]

Mobilizing American democracy for war, Herring argued, confronted obstacles inherent in its liberal tradition. The Jeffersonian ideal of the citizen springing to arms in a war emergency gave rise to an "utterly indifferent" response to preparedness and national security organization. Herring maintained that interest groups, guided by their own needs and entangled in local and congressional politics, fragmented the country. This pluralistic system lacked even basic balancing mechanisms for institutional continuity. No organization blended interdependent parts into a whole or educated the public to the need for preparedness and national security. There was nothing to promote continuity and consistency in national policy. "There is in our system," Herring concluded, "no gyroscopic device which ensures that the foreign policy of a representative government will maintain a constant level regardless of internal political forces."[18]

Democracy required expert guidance and orderly organization to combat totalitarian forces and to avoid socialism. So Herring searched among European and American models for a "gyroscopic device" to balance the diverse interests and fragments into what he called "a national symphony." Like Carl Friedrich, William Yandell Elliott, and other close colleagues, Herring felt that European corporatist structures such as German Federal Economic Councils, the French Centrist Party, the Czechoslovak Economic Advisory Council, and the British Labour Cabinet offered possible organizational solutions. They charted a path between laissez-faire competition and totalitarianism. These corporatist structures sought to balance political and economic interests, provide unity, standardize, and discipline, while protecting individual free enterprise and building mechanisms to link national resources to security. Americans experimented with these forms only when a war emergency forced what Herring called a period of democratic mobilization. Then, they built temporary structures such as the War Industries Board of World War I and the National Defense Advisory Commission of 1940. As soon as the crisis passed, they abandoned the emergency agencies.[19]

Herring compared the two periods of democratic mobilization and insisted that a profound change had occurred between 1917 and 1941 that made temporary national security structures obsolete. A "colossal world revolution" had given rise to totalitarian systems and ideologies and to unprecedented innovations in the science and technology of war. Distinctions between peace and war had disappeared, replaced by a constant state of crisis. New conditions demanded permanent institutions to coordinate military, diplomatic, and economic policies for national security. That machinery, Herring concluded, must be run by an administrative elite in government and by patriotic experts from the private sector through quasi-official advisory councils, commissions, and associations. Herring exemplified his own ideal by serving as a consultant to the Bureau of the Budget, War and Navy departments, and now with the Eberstadt study group. The latter provided the ultimate opportunity [to influence public policy] because Herring's concepts permeated the final Eberstadt report. When Congress published the report in late 1945 and then used it as a framework for national security legislation, Herring's synthesis became part of public policy.[20]

Soon after Herring's arrival, he was making an impact on the study group. He warned that twelve of Eberstadt's twenty-nine researchers were veterans of the ANMB and WPB and focused solely on logistical problems, making the scope of the study too narrow. After the dropping of the atomic bomb on Japan, Herring called for attachment of scientific experts to the committee. "If we assume that great mass armies and large fleet operations have been rendered obsolete by the atomic bomb," he explained to Eberstadt, "the logistical problems of the future will be very different from World War II." He posed the question: "How have scientific advances changed the nature of warfare?" Though Eberstadt doubted there had been a significant change in that area, he agreed to attach more scientific and technical experts to the group. New members included Philip M. Morse, an MIT engineer and director of the Navy Operations Research Group at Brookhaven, and George M. Snellings, a radio and radar communications expert. Simultaneously, Eberstadt secured an advance copy of Vannevar Bush's "Science, the Endless Frontier" and began interviewing Bush, James Conant, Jerome Hunsaker, and other members of the scientific elite who worked on government projects, particularly the atomic bomb project.[21]

Eberstadt recruited an aviation expert as well. He knew that King, Forrestal, and naval aviators opposed a separate air force. Interviews with Paul Nitze, a Dillon, Read executive who worked on the strategic bombing survey; Grumman executive Jake Swirbul; Eugene Wilson of Pratt & Whitney Aircraft; and T. E. Braniff, on whose board of directors Eberstadt sat, had him convinced that the future of American commercial and defense policy lay more in the air than on the sea. "In the past it is true that the Navy has been our first line of defense and has received the first shock of attack," Eberstadt wrote, "but this may not be so in the future." A separate air force department made sense in that it reflected changes in technology, strategy, and warfare. He attached commercial aviation pioneer Nicholas Ludington to the group. Founder of a Camden-Philadelphia airline, which later merged with United Airlines, Ludington had served as Forrestal's special assistant for contracts and supplies in the Bureau of Aeronautics, and, manifesting strength of connections, had married Harry Hopkins's sister. Thus, for a number of reasons Eberstadt expected Forrestal to listen to Ludington's recommendation for a separate air force department in the national security organization.[22]

Eberstadt also accepted the advice of Bert Fox; Adm. Russell Willson, JCS intelligence expert; and James Phinney Baxter, Williams College president, to prepare a special section on a central intelligence agency. Eberstadt had long considered the collection of financial and political intelligence to be a central part of the domestic and foreign investment business. He shared Forrestal's view that in executive management and administration, whether in business or government, the accumulation of data was a valuable policy tool. Of all the postwar structures, Eberstadt told Baruch, "a Central Information Bureau" was the most important. Probably he preferred the help of longtime friend William Donovan of OSS, who he placed on his most intimate Good Man List along with David Sarnoff, Averell Harriman, Herbert Bayard Swope, and Walter Lippmann, but settled for Sidney Souers, assistant director of Naval Intelligence and later head of the Central Intelligence Group and secretary of the National Security Council.[23]

During the summer of 1945, Eberstadt's experts pursued their individual studies, meeting from time to time with their boss to iron out basic concepts, structures, and problems. They rarely sat as one committee. "I met the rest of Eberstadt's group only twice at a lunch," science expert Philip Morse recalled, "and wrote my appendix to his report separately, with very little contact with the rest." When Eberstadt did discuss their research, recalled group member Myron Gilmore, he was arrogant about his own opinions. Robert Howe Connery disagreed, claiming that without exception members of the group spoke of Eberstadt's tolerance of ideas, patience during debate, and mental keenness in absorbing and synthesizing information. "I know of no one in the naval service who succeeded so well in winning the wholehearted loyalty and admiration of his staff," Connery wrote Eberstadt. As one member observed, when the group submitted research, Eberstadt "then withdrew and made up his own mind about what could and should be said without much regard to what the Commission was doing."[24]

Nevertheless, the inner group consisting of E. Pendleton Herring, C. Douglas Dillon, Eugene Catron, Maxwell Rabb, Edward J. Carr, and Edward Hidalgo, a Columbia law graduate and member of Eberstadt's old law firm, worked together. They interviewed leading wartime figures in naval mobilization, particularly representatives from prime wartime contractors RCA, General Electric, General Motors, U.S. Steel, and United Aircraft (Pratt & Whitney). They spoke extensively with executives and staff members from the Army Navy Munitions Board, War Production Board, War Manpower Commission, and Office of War Mobilization and Reconversion. They focused on CMP veterans. They solicited opinions on postwar national security from key congressional leaders, major journalists, and from organized labor and agriculture in line with both Eberstadt's and Herring's corporatist approaches to balancing major sectors and interests in a national security program.

Predictably, data reflected the opinions of a coherent collection of naval mobilizers and the type of corporate organizers who were on the Good Man List. Eberstadt's group talked to few who held the single-department perspective on national security organization or a New Deal view of government. They interviewed only Harold Smith, budget director; Robert Lovett, assistant secretary for air; army aviator F. Trubee Davison; and a few minor War Department civilian mobilizers. They considered an interview with General Palmer "to discuss our ideas and see how effectively they stand up against his criticism" but never talked with General Marshall's intimate advisor on the single-department plan. The heavy emphasis on Navy Department witnesses was not entirely Eberstadt's choice; the War Department blocked interviews until Forrestal intervened directly. "It is, of course, difficult to arrive at an objective conclusion if the proponents of differing views have reserve about expressing them," Forrestal wrote the secretary of war. When Eberstadt received permission finally "to talk to anyone," it was too late. He had completed the study.[25]

Eberstadt's conceptual approach, which differed markedly from the War Department's scheme, took shape between July and early August. After talking with Woodrum and reading the transcript of the Woodrum Committee Hearings, Eberstadt concluded that his study group's mandate lay in identifying gaps between primary structures and support bodies in the current organization and in building mechanisms to close those gaps. There were critical gaps between army and navy, foreign and military policy, strategic planning and logistic implementation, military and civilian mobilization agencies, executive and legislative branches, and between government and business. Such gaps created competition for power and resources, a situation that Eberstadt felt paralleled American political-economic conditions. While the economy suffered cycles of inflation and depression that required stabilizing by something like Herring's "gyroscope," so too national security faltered with its own cycle of instability. "We have always gone from extreme and unwise parsimony in time of peace to unbridled and equally unwise expenditures in time of war," Eberstadt observed.[26]

Eberstadt's solution for closing the national security gaps lay, as it did in his corporate world, in finding the right organizational balance through reorganization and structural innovation. "The great scope and complexity calls for constant reorganization and regrouping," Eberstadt noted "so that organizations may be in harmony with the main stream of problems, so that they may be furnished the fullest information, may reach a balanced judgment, and may effect a decentralized execution of their decisions." Primary organizations had grown so large that a new set of coordinating structures had to be created to ensure linkage between national resources and national security. Eberstadt believed that coordinate units, rather than a single department of defense, would provide better security and do so without crushing the competition and productivity of the democratic system that were necessary to assume wider global obligations. Moreover, intermediary organizations would give business experts continued access and influence.[27]

As Eberstadt designed this system of coordinated national security units, he was interrupted by a presidential job offer. In June and again in August, 1945, Truman asked Eberstadt to head the United Nations Relief and Rehabilitation Administration program for displaced persons in Germany. On the surface, this seemed an effort by the main proponent of the single-department concept to get Eberstadt out of the country and isolate Forrestal. On closer examination, though, it seemed that the impulse to recruit Eberstadt resided with members of a prewar preparedness and internationalist organization known as the Century Group, all of whom were on Eberstadt's Good Man List and now served in various capacities in the Truman administration. Will Clayton, Allen Dulles, and Lewis Douglas, perhaps impressed with Eberstadt's idea of a peacetime Lend-Lease program for Europe, saw in their friend the right type of assistant to aid General Clay, David Sarnoff, and others in guiding German economic recovery as a bulwark against Soviet influence. In any case, Eberstadt refused the job. "I don't like to establish a practice of saying 'No' to Presidents," he told Forrestal, "[but] I have become very much interested in your job and you will see from the tentative draft outline that I left with you, it is just getting under way and I would hate to drop it at this point."[28]

Eberstadt wanted to survey select members of what could be called a loosely connected policy elite. These individuals were similar to those in the Century Club and its larger organization, the Council of Foreign Relations; to members of the Business Advisory Council; and to representatives of the newly formed Committee for Economic Development (CED). Eberstadt felt that they could give substance to his national security recommendations. Herring advised that Eberstadt also contact leading academics such as Edward S. Mason, dean of the Harvard Graduate School of Public Administration. Eberstadt prepared a list of such names, cleared every one with

Forrestal, and sent a survey letter to everyone on the list, asking for their views on unification and national security organization. Those national leaders who replied included the scientists Vannevar Bush and James Conant, J. F. Dulles, John Hancock, journalists Hanson Baldwin and Arthur Krock, and PACH director Herbert Emmerich. Also answering Eberstadt's inquiry were GM executive O. E. Hunt; Paul G. Hoffman, Studebaker executive and BAC-CED organizer; labor leader William Green; diplomat Sumner Welles; Hugh Drum, former army officer and current head of the Empire Corporation; William V. Pratt, the former navy CNO and currently a *Newsweek* executive; Eric A. Johnston, of the Chamber of Commerce; Chicago banker Edward E. Brown; historian Charles Beard; and a number of corporate lawyers and business leaders close to Eberstadt.[29]

The unification study provided an opportunity for this select group of respondents to discuss wider political, economic, and strategic issues. Their letters formed a coherent body of ideas for postwar national security. Similarities in background, corporate connections, opinions, and experiences of those surveyed, as well as the nature of specific organizational questions asked, accounted partly for this unified response. Despite the pre-set conditions, the passion and force with which this elite group articulated their common vision of postwar problems and organization in the summer of 1945 helped explain the final shape that national security assumed in the immediate postwar period. All agreed that after the war the United States would have to assume major peacekeeping functions, rebuild the world economy, and ensure international stability in the face of revolutionary unrest. They forecast a worldwide struggle between American democratic capitalism and Soviet communism. In August of 1945 many of these respondents were already predicting another war. "The preparation required now for the next war," wrote Fred Searls, OWM administrator, "is to protect our currency, balance the budget, [see to] repayment of the debt, maintain small development programs, and recover healthy national economy for the next war effort."[30]

Expectation of continuing crisis and confrontation necessitated—or justified—in the minds of those surveyed the construction of a permanent network of interconnected formal and informal structures as part of a national security establishment. "I think the civilian war-machine should be maintained at all times in skeleton form," advised WIB veteran John Hancock. He offered a decidedly corporatist perspective on national security and therefore recommended that interorganizational and coordinating mechanisms close the gaps between powerful sectors and interest groups, eliminate national political-economic conflicts, provide government-business partnership, and promote a consistent national security program. At the same time, Vannevar Bush advised that such national security machinery include a permanent war cabinet and an elite general staff complete with its own unique uniforms. John Rockefeller called for institutionalization of the State-War-Navy Coordinating Committee as a Cabinet-level body, while Rockefeller executive Langbourne Williams, Jr., of the Freeport Sulphur Company offered the idea of a permanent national security council with nine members selected by an independent nominating committee representing industry, labor, and other national interests. "To protect the work of the Council against temporary displeasure of the Congress," Williams continued, "it should be provided that its appropriations could be added to by gifts from private foundations or individuals, with appropriate safeguards against too much being contributed from one source."[31]

All of the individuals surveyed wanted an organizational elite that stood above local interests and politics to run this permanent war machine. "The working of even the best administrative machinery," Elihu Root, Jr., reminded Eberstadt, "will depend on the quality of the men who run it." The scientists recommended a board of eminent scientists, the business respondents advocated a business and labor advisory council, and PACH director Herbert Emmerich thought that public administration experts should manage the war machine. Emmerich indicated that the idea of civilian control over the military was his rationale for packing the national security establishment with "civilian aids and the means for control through budget and personnel offices and other devices." Gen. Leslie Groves, military head of the Manhattan Project, informed Eberstadt that he preferred "true military statesmen" to run the national security machine, while Williams avoided classifying the ideal national security officials as civilian or military so long as they were "some of the ablest minds in the country." Williams wanted these able minds to study military and foreign policy and prepare advisory reports like those of the Royal Commissions in England.[32]

Some stressed the impact of scientific and technological developments on national security. Hanson Baldwin, *New York Times* military editor, explained to Eberstadt that the invention of "Buck Rogers weapons" posed an unprecedented threat to American security. "What we now know of the capabilities of rockets it seems quite probable that within the foreseeable future, man will be using transoceanic rockets perhaps with Atomic warheads," Baldwin predicted in August, 1945. David Sarnoff, RCA executive and longtime Eberstadt comrade, insisted that in three years the Soviets would have atomic bombs with the means to deliver them. He warned that American scientific, industrial, military, and communications sectors had to unite in their vigilance against a surprise atomic attack. United Aircraft president Eugene E. Wilson advised Eberstadt that there must be intimate collaboration between government and private industry for the research and development of new weapons. He wanted the government to subsidize commercial airlines and research on new aircraft as the best path to security and prosperity. "Technical leadership so vital to keeping peace between ourselves and Russia with two fundamentally opposed ideologies," he told Eberstadt, "can thus be an investment rather than a burden on the public purse."[33]

By early September, Eberstadt had received forty letters on unification and national security organization from his carefully selected group and had completed interviews with several dozen others, including Nelson and John Rockefeller, Edward Stettinius, Harry Hopkins, Owen Young, Herbert Hoover, and Bernard Baruch. Baruch read and revised a draft of the report. Because the opinions of these men reinforced his own corporatist impulse, Eberstadt employed their theories throughout his report. He also borrowed from other organizational influences, including the traditional idea of joint army-navy boards, Progressive Era public administration theory, and business and industrial management concepts, but in the end advice from the interviews and letters weighed more heavily because it most closely approached Eberstadt's own corporatist education.

The final version of the Eberstadt Report on unification of the armed services began with some of the generalizations found at the base of all American corporatist thought (from that of Wilson to Hoover, and from Baruch to Eisenhower): that solutions to the various problems of modern industrial society lay in large-scale corporate organization and the development of coordinating administrative and managerial mechanisms. Eberstadt explained that the national defense establishment had become a complex, unwieldy organization akin to the great industries that had transformed modern America. "The tremendous growth of our military establishment challenges the skill of the most able organizers," he explained. "Thousands of functions and millions of material items are involved." Corporate structures run by skilled managers had rationalized big industry and business and thus could also rationalize the huge defense bureaucracy and make it more responsive to business controls.[34]

Instead of Marshall's unitary system of national security that focused on governmental controls, Eberstadt recommended adoption of a corporatist system of coordinating agencies. He noted that this type of organization would allow the nation to derive the "tremendous benefits that arise from the parallel, competitive, and sometimes conflicting efforts which our system permits." Though his report implied the construction of an autonomous and parallel national security government, Eberstadt insisted that the corporatist form of unification provided constant military preparedness within the "framework of our democratic system of government." A series of interlinking bodies rather than consolidation into a central organization protected the domestic economy and industrial production. The unitary principle seemed to be economically unsound and financially destructive. "It cannot be permitted," Eberstadt wrote, "to cause price, wage, and other financial maladjustments, which through impairing public morale, would have the same effect."[35]

Eberstadt's unification blueprint was grounded in yet another corporatist strategy: institutional arbitration and public-private power sharing. All the agencies and structures that he suggested in his report encouraged movement of skilled corporate managers between government and the private sector. The report advocated having advisory committees to handle mobilization planning, with representatives of business, industry, labor, and agriculture to maintain contact with the national security agencies. Almost as an afterthought, Eberstadt noted that the defense agency must also develop a continuous association with congressional committees.[36]

Another corporatist aspect of Eberstadt's report lay in its provisions for guidance of coordinating mechanisms by an organizational elite, a "small group of people" like those that Eberstadt saw in the British Committee on Imperial Defence. "In their corporate capacity," he explained, "they have served as the secretariat of both the Defense Committee and Cabinet." This "small group of people" (or cluster of groups) managed national security and integrated foreign, military, and economic policy. Competent men in positions of authority, Eberstadt continued, needed coordinating machinery to establish collaboration between the various elements of the national security system. "Separate departments provide a greater representation of specialized knowledge; they provide a greater aggregation of experienced judgment and insure representation of varying viewpoints," he wrote. A single secretary in a merger of army and navy into one structure, as proposed by Marshall and Truman, Eberstadt warned, might become a puppet of the military establishment. "There is also the chance that the civilian and military chiefs might be tempted to exercise prerogatives reserved by the Constitution to the President," he wrote.[37]

And finally, Eberstadt's unification report addressed international as well as domestic issues and assumptions. It suggested that national security required cooperation between nations and the promotion of international stability through American financial and economic guidance, rather than by military power. Eberstadt's report contained the type of Wilsonian internationalist strategy that historian Thomas J. McCormick labeled "a system of globalized corporatism." Eberstadt insisted that true security stemmed from America's ability to use its productive strength to prevent unrest in the world by stabilizing the international economy. "True economy lies in constant and diligent attention to our national security to avoid the causes of war," Eberstadt explained.[38]

Eberstadt's plan would create paths through which business could dominate national security. It recommended formal cooperation between the state and major economic power blocs. Eberstadt, either consciously or automatically drawing from his corporate background, sought to build corporatist structures through unification. If corporatization afforded economic stability, subordinated individual interests to national unity, minimized conflicts, and brought Good Men to the top, then it might offer the perfect alternative blueprint for national security organization that Forrestal needed to counteract the single-department scheme.

National security operated in the larger political-economic environment. An indifferent, poorly informed public allowed destructive competition between civilian and military to continue because nothing threatened the national security. In the past, the United States had had time and resources to mobilize the nation and defeat the enemy. Now, Eberstadt observed, "the margin of resources is no longer so great while the scope of the enemy's weapons is far greater."[39]

Each structure recommended in Eberstadt's final report (most of which appeared two years later in the National Security Act of 1947) facilitated institutional arbitration between the public and the private sector. The National Security Council (NSC) became the keystone of Eberstadt's coordinate system. It stood as "a permanent vehicle for maintaining active, close, and continuous contact between the departments and agencies of government responsible for our foreign and military policies, including formulation and coordination of such policies, the assessing and appraising of our foreign objectives, commitments and risks, and keeping these in balance with our military power." Baruch saw the NSC as the essence of a corporatist strategy because it created "a select executive body charged with responsibility for America's global strategy."[40]

The proposed National Security Resources Board (NSRB) became a basic mechanism to balance the nation's supply of resources with its military demands. The board kept a peacetime resource inventory and made the "national balance sheet of resources solvent through supervising disposal of present surpluses, advocating sound policies of conservation, guarding against shortages in strategic materials, and developing and maintaining adequate information on the manpower, resources, and productive facilities of the nation." An inventory of the resources of foreign nations resided in yet a third structure, the Central Intelligence Agency (CIA). The CIA provided that vital upward flow of information to the top corporate directors, a basic principle to those who ran modern business organizations. Other interlinking mechanisms included the Military Munitions Board to standardize contracting, design, and production techniques; the Research and Development Board to promote close cooperation between civilian scientists, industry, and the government; and the Military Education and Training Board to instill "a mental attitude of cooperation" throughout the nation. Recommendations for a permanent statutory JCS and independent Air Force department, co-equal with the separate War and Navy departments and headed by a civilian secretary, rounded out the report's structural innovations.[41]

None of these forms was especially radical. All had precedents in temporary organizations during both world wars, especially in the WIB and WPB organizations. The independent air force had been advocated and debated vigorously for thirty years. A munitions board had existed since 1920. The report took pains to preserve the autonomy and integrity of the separate services and to make overall structure conform to the traditions, customs, and constitutional parameters of the United States. Echoing the Morrow Report of 1925, it recognized the special functions of sea and land services that required separate institutions. It insisted that separate organizations engendered healthy competition and high morale. At the same time, the development of permanent coordinating machinery to provide "sustained and continuous defensive action" and to integrate foreign, military, and economic policies for global obligations was unprecedented. Eberstadt suggested innovative structures to adjust national security design to the "revolutionary factors of speed and destruction" that marked modern war.[42]

When Eberstadt submitted his final unification report to Forrestal on September 25, 1945, it brought the immediate postwar national security debate to its highest level of intensity and most crucial organizational stage.

7. Politics of the Eberstadt Report, 1945–1946

The farewell luncheon for Eberstadt's Unification Study Group, held amid the victory celebrations of September, 1945, drew the most emotional display by Forrestal since his arrival in Washington five years before. The outpouring arose both from the end of the intense war experience and a growing concern that America remain armed, strong, and vigilant in the face of an uncertain and dangerous future. Forrestal stood at the head table, placed his arm around Eberstadt, and called for silence so that they could look each unification team member in the face. He wanted always to remember the moment. That image of "the strong emotional bond between the two men," remained vivid in the memory of Wallace MacGregor thirty-five years later. "Needless to say," MacGregor wrote, "one can hardly forget a day when the Secretary of the Navy would pass up a Victory celebration of a war which worried him greatly to meet with 30 officers and his best friend as an expression of thanks for the effort to produce a plan for him which he believed in so sincerely."[1]

This bond and Forrestal's commitment to Eberstadt's report on unification and postwar national security determined the character and course of the politics of national security that followed. Excerpts of the top-secret report leaked out at once–most likely via Forrestal himself–and made their way through the bureaucratic maze of Washington. AAF planners dissected it, War Department planners brought in Generals J. Lawton Collins and Dwight Eisenhower to testify against it, and Navy planners reduced it to brochure form and circulated it as the "Navy Plan for National Security." Printed by the Senate Naval Affairs Committee (SNAC) on October 22 (although not declassified until December 17), underlined copies of Eberstadt's report appeared on the desks of congressional leaders as they considered unification legislation. Newspaper editorials and radio commentators analyzed it. The Navy League, Navy Industrial Association, and other interest groups promoted it. White House assistants summarized and debated it, while President Truman modified his own reorganization design to counter specific recommendations in the Eberstadt Plan. Everywhere they turned, budget director Harold Smith told the president, they confronted what he saw (probably with good reason) as a dangerous threat from the Eberstadt Plan to presidential authority and democratic institutions.[2]

Still, the report's influence and contribution depended on Forrestal's commitment to it, and that commitment derived largely from his personal bond and partnership with Eberstadt. Between September, 1945, and early 1946, Eberstadt drove his friend to present the Eberstadt Plan as the blueprint for postwar American national security organization. Forrestal accepted this advice with increasing regularity. After years of public service and continual crisis, the secretary had begun to show what some noticed as signs of weariness and strain. In a way, Forrestal recognized his own decline when he told Arthur Krock in October, 1945, about "the corrosion that is apt to accompany long continued public service." Forrestal leaned more heavily than ever on Eberstadt, dispatching a government plane to New York to summon him on a moment's notice. He asked Eberstadt to cut short a trip to Mexico, pulling him away from his seriously ill daughter. He intervened with the New York Supreme Court to excuse Eberstadt from jury duty so that he could be by Forrestal's side for consultation during key periods of the unification crisis. Eberstadt's advice was always the same: stand behind the Eberstadt Report 100 percent. If he did, Eberstadt explained, Forrestal could not lose the battle for national security organization and policy.[3]

The Eberstadt Plan meant different things to each partner, although their common motives, purposes, and actions obscured those differences. Forrestal saw the report first and foremost as a bureaucratic tool with which to counteract single-department politics and strengthen the navy's position as the first line of sea-air defense. He favored the report as well, Forrestal explained to Winston Churchill, because it reflected the ideal model of the British war cabinet system outlined in Sir Maurice Hankey's writings. "I told him that Mr. Eberstadt had made a study for the Navy of the whole subject of defense," wrote Forrestal, "and that the germinal basis of his plan was the minute by Lord Hankey on the operations of the Imperial General Staff in England."[4]

At the very least, Forrestal hoped to use the report's central thesis of coordinate structures in an effort to institutionalize the Joint Chiefs of Staff, the State-War-Navy Coordinating Committee, and other wartime joint bodies on a permanent basis to link foreign and military policy. Forrestal also liked the corporate organizational implications that paralleled modern business organization with its centralized policymaking, decentralized operation, and balancing and forecasting of resources and supply needed to carry out national objectives. In his mind unification revolved more around retaining a type of military organization than espousing a corporate philosophy. Partly because Navy planners failed to develop a postwar scheme for

Forrestal to use against air power proponents in the unification struggle, he turned to the Eberstadt Report as "an invaluable point of reference."[5]

On the other hand, Eberstadt conceived his report as more than a device to save the navy, promote a sea-air power doctrine, or build a national defense establishment for the postwar period. The Eberstadt Report was a model to begin shaping a larger postwar political-economic organization and business-government cooperation. The military question was far less important in his equation. "Our military power stands in the middle between foreign policy on the one side and our national resources, human, spiritual and material, on the other," he explained.[6]

Forrestal excitedly revealed the Eberstadt Plan to *New York Times* publisher Arthur Sulzberger, a proponent of the single-department plan. He told Sulzberger that Eberstadt had created an alternative to the War Department scheme that would change Sulzberger's mind. He asked the publisher to keep the news in strictest confidence. However, Forrestal chafed to get the report out in the open through a press release or as articles by "legitimate authorities" such as journalists Hanson Baldwin, Walter Lippmann, or G. F. Eliot. Eberstadt, too, wanted the report made public so that he could discuss it with "interested people." He pressed Forrestal to persuade the White House to remove the top secret classification, and the secretary made the request a few days later. Forrestal proceeded cautiously, though. "Obviously, I would not want to do it," he explained to White House counsel Sam Rosenman, "if it is not in line with the wishes of the President."[7]

At this juncture, President Truman's wishes and his relationship with Forrestal began to influence the course that the navy secretary followed and to determine his behavior nearly as much as his bond with Eberstadt did. Though neither entirely trusted the other nor established close rapport, Truman and Forrestal developed a symbiotic arrangement that compelled intimate association for four years of struggle over unification. A Forrestal biographer claimed that the relationship with Truman revolved around Forrestal's identification of the president as his father, an authority figure that he both loved and hated. More likely, the two feisty men needed each other. Truman wanted the popular and seemingly strong and uncomplicated secretary in his Cabinet to adjust the navy to administration policy, particularly budget-cutting, demobilization, and security plans. He expected the secretary to rein in the admirals and to link the administration to Carl Vinson, David Walsh, Leverett Saltonstall, Styles Bridges, and other navy (and Forrestal) friends in Congress. In addition, Forrestal defended and advanced the president's own hardening policy line toward the Soviet Union. Forrestal praised the president for the way that he "stood up to the Russians." Unlike Commerce Secretary Henry Wallace and Interior Secretary Harold Ickes, two Roosevelt holdovers that Truman subsequently fired, Truman trusted Forrestal to develop and back (although not to lead) the administration's emerging Cold War policy.[8]

For the remainder of 1945, the president ignored or chose not to respond to evidence that Forrestal's stubborn opposition to administration defense reorganization plans bordered on insubordination. He dismissed claims by Wallace and Acheson that Forrestal was the "lying leaker" in the Cabinet. Possibly, Truman allowed Forrestal to cause friction so as to demonstrate indisputably the need to unify the armed forces into a single department. Then again, Truman used administrative politics just as much as did Forrestal, Eberstadt, or the organizational managers from the Good Man lists. He employed reorganization of government to achieve larger goals of strengthening the New Deal presidency and in promoting regulatory agencies. Moreover, Truman considered universal military training more vital than unification of the military establishment. He never viewed the controversy as a crisis as did Forrestal. This explained Truman's toleration (and probably exploitation) of the navy secretary's opposition to the single-department approach. In any case, Truman gave Forrestal wide latitude in defying the administration's announced position, and Forrestal's subsequent activities more than those of any other participant in the politics of unification between 1945 and 1947 determined the final form for postwar national security organization.[9]

Regardless of whether Truman was engaging in the calculated political tactics of a strong executive or displaying what one scholar has called Truman's timorous quality when confronted with difficult problems, the president encouraged the American defense establishment to break into feuding and demoralized factions at a critical moment in the formulation of postwar policies toward Western Europe, the Soviet Union, and East Asia. While he debated whether to adopt internationalist accommodation toward the Soviets or a policy of confrontation and containment, Truman saw the American military plunge into the worst interservice conflict in its history. This controversy stemmed partly from his own caution. Truman claimed later that he pushed for unification from his first day in office, but in fact he took no direct action for months, despite pressure from the budget director and the War Department to make a public statement. Though the White House prepared a unification brief, including an analysis of the Eberstadt Report, by mid-October, 1945, staff member George Elsey, Jr., noted that "the president's interest is passive at present." Truman told Smith that other matters took precedence and that he had no plans to introduce unification until well after the first of the year.[10]

Truman's delay was understandable from the perspective of a period that the president called the most hectic in history, when leading the nation was like riding a wild tiger without being devoured. FDR had left Truman unprepared for war and postwar leadership; he had not discussed strategy, policy, or even developments such as the atomic bomb with his vice-president. He had not made known any firm opinion on defense reorganization, an autonomous air force, or unification. Truman's search of White House files uncovered no references to the single department. Admiral Leahy, who along with Marshall provided the most immediate counsel and made the greatest impact on Truman's thinking about military matters, insisted that the only opinion that the late president had expressed was that he intended to retain naval aviation and the marines as effective parts of the navy after the war. Pledged to continue FDR's course and follow a smooth transition, Truman decided not to introduce at the moment plans never approved by President Roosevelt.[11]

Truman expected the Joint Chiefs of Staff and top military and civilian advisors to iron out interservice differences and recommend measures to eliminate waste and duplication. He ordered them to make economical estimates of postwar force levels and budgets. "The time had come to put an end to piecemeal legislation and separate planning for the services," Truman insisted. He wanted the JCS to reach an agreement on a policy for overseas bases, demobilization, air force strength, and the definition of service roles and missions. He expected the Joint Chiefs to recommend as the basis for postwar military organization the structure outlined in the Richardson (JCS Special Committee) Report. However, deep divisions on the JCS between Marshall and AAF chief "Hap" Arnold on one side and Admirals King and Leahy on the other prevented agreement. Instead of solving the problem, the JCS submitted the Richardson Report to Truman on October 16 with four separate options for postwar defense organization. The JCS advised that the president either employ his constitutional authority to appoint a single defense secretary, retain the existing organization, name a board of experts to make recommendations, or leave the entire question to Congress.[12]

Truman favored the last recommendation. He considered Congress the "responsible commission" to reorganize national defense through legislative action. The naval and military affairs committees of both houses provided the forums to hear arguments from different sides—including the Marshall, JCS Special Committee, and Eberstadt plans—and to find a compromise and draw up a new national defense law. Truman suspected that some disloyal legislators dominated by local selfish interests might not follow his leadership; and indeed the Vinson-Walsh Concurrent Resolution that fixed the number of ships for the postwar navy without presidential review seemed to Truman "a dirty trick on the part of the sponsors." He had faith in the congressional politics and process that had been the central focus of his career. "We may have to use a little old-fashioned ward politics here, to get some action in the national interest," he told Harold Smith at the budget office.[13]

Truman let the Senate Military Affairs Committee (SMAC) become the focus of defense reorganization and unification debates, particularly since from his perspective the early sessions went smoothly on hearings that began in mid-October for a bill to unify the armed forces in one department. Expecting "a knockdown, drag-out fight in Congress over the question of a single department," Marshall's staff had far better prepared the War Department for congressional hearings than it had for the Woodrum Committee sessions. A War Department team of Lieutenant Generals J. Lawton Collins, William F. Tompkins, and Harold L. George briefed witnesses and directed historical research. They consulted with the bill's sponsor and leading SMAC member, Sen. Lister Hill of Alabama. Marshall ordered Floyd Parks, Eisenhower's public relations expert, rushed from Supreme Allied Headquarters in Europe to assist in the army's presentation. The result was a steady stream of effective testimony by War Department witnesses buttressed by carefully crafted charts and tables that defended unification of army and navy into a single department of national defense. When Forrestal introduced the Eberstadt Report to the committee, General Collins appeared a few days later and discredited it with his own plan (actually the Marshall Plan), an effective display of organizational charts, and the explanation that Eberstadt's coordinate system meant three military departments rather than one and therefore promoted militarism.[14]

The most effective testimony was that of Eisenhower, who swept the packed Senate chamber with his charm and image as an honest, frugal soldier intent upon promoting team spirit and team play through unification of the military. Perhaps the most popular military figure in the country, and just returned from Supreme Allied Command preparatory to replacing Marshall as chief of staff, Eisenhower basked in the publicity of the hearings. "Klieg lights were on the whole time," Willett wrote Eberstadt after attending the session, "four newsreel movie men were grinding away at intervals and sound reels were also occasionally taken." The single-department plan, Eisenhower assured the senators and press, presented a perfect organization for civilian-military and army–navy–air force teamwork. "Unification would give us an effective fighting force with 75% of the number of personnel that would be required if we had separate departments," he predicted.[15]

AAF witnesses, who saw support of unification and the single department as the best path toward air force autonomy, added their own dynamic and often spiteful anti-navy testimony that portrayed the navy and separate service organizations as outdated and a threat to national security. James Doolittle, who early in the war had launched a bomber attack against Tokyo from the deck of a big aircraft carrier, now attacked that same vessel as a useless, obsolete platform vulnerable to air attack. Doolittle brought his own set of charts and maps that showed the senators how AAF bombers could reach deep interior land targets, like those in the Soviet Union, beyond the limited range of carrier-borne aircraft. Encouraged by Senator Hill, Doolittle criticized the navy, declaring that his deceased B-29 boys rested uneasily in their graves when the navy claimed that sea power had won the war against Japan. A few days later, Gen. Carl A. ("Tooey") Spaatz, soon to succeed Hap Arnold as AAF commanding general, contradicted navy assumptions that sea power provided the first line of American national defense. Spaatz explained that the supersonic jet, guided missile, and atomic bomb had made the entire United States as vulnerable to aerial attack as London had been during the last war. "The Pearl Harbor of a future war might well be Chicago, or Detroit, or Pittsburgh, or even Washington," Spaatz warned. Only an independent U.S. Air Force offered future security against such attack.[16]

The president welcomed AAF comments. For Truman, the air generals built much needed "air mindedness" in the nation and public support for air power, an air force, and unification. During the hearings, AAF interests consulted with the White House. William P. Redding, a National Aeronautic Association executive who had collaborated earlier with Truman to pass a Civil Aeronautics Act, explained that he would work "behind the scenes" to organize grass-roots support for the president's reorganization plans and pass on information "in the strictest confidence" to the White House. Redding combined scattered aviation groups behind Truman, including the Air Power League, Air Force Association, Wings Clubs, Aeronautical Training Society, and the Civil Air Patrol. More important, General Spaatz served on the National Aeronautic Association's strategy board as permanent AAF liaison. Redding wrote Truman, a life member of the association, that based on their earlier confidential conversation, he wanted to develop ties to "someone in the White House organization to whom we could report on progress of the campaign for the single department of defense for co-equal status for sea, land, and air."[17]

If Truman expected the hearings to provide the platform to promote air power and unification, he was disappointed. Despite the best efforts of Senator Hill and War Department witnesses to build a strong case for the single-department format, the hearings opened the way for Forrestal, Eberstadt, and Assistant Secretary of the Navy Hensel to present their plan as a more reasonable and comprehensive alternative to the army's narrow approach to national security organization. Gradually the senators and press focused on Eberstadt's report rather than the single-department blueprint favored by the president. By mid-November, one army witness admitted that parts of Eberstadt's study were the finest thing that he had ever seen on the subject. An AAF special planner who analyzed the report explained that it had brought the debate to a much higher level than the War Department plan. Journalist George Fielding Eliot appeared still more enthusiastic, saying that the Eberstadt Report was "the best and most clearly reasoned proposal on the systematic control of our foreign and military policies" that he had ever seen. Members of the SMAC focused on the report, too. They called it the only impartial, complete, and professional study of national security. They respected Eberstadt's approach, Maxwell Rabb explained, telling Eberstadt that he appeared "as anything but a Navy stooge."[18]

Instead of solidifying Truman's political support for reorganization, the Senate Military Affairs Committee hearings had provided a wedge for Eberstadt to influence not only Navy Department politics but also congressional, White House, and War Department bureaucratic maneuverings. As Eberstadt's report gained support, the president's approach lost stock. Rather than a master of congressional politics, Truman emerged—at least over unification—as a chief executive forced to react to, and eventually compromise with, the Eberstadt Plan for National Security and in the end to accept structures, concepts, and solutions that reflected Eberstadt's vision rather than his own.

The War Department's assault, particularly the AAF tirades, shocked the Navy Department and Forrestal into adopting the Eberstadt Report as the "Navy Plan for National Security." Such a course had not been certain when Eberstadt had submitted his report in September. After the SMAC hearings opened a month later, neither Forrestal nor navy witnesses presented the report as their official position. They found recommendations for an autonomous air force unacceptable, and Forrestal rejected Eberstadt's suggestion that they employ the report as a basis for compromise and agreement with Robert Lovett, assistant secretary of war for air and a close personal friend of both. Forrestal doubted that Lovett could bring the army into line behind the Eberstadt Plan. "I am giving a good deal of thought to your conversation with Lovett," Forrestal wrote in early October, continuing, "I am entirely confident that we would have their complete support on the thesis of a department for Air, but I must say that I haven't quite reached that conclusion."[19]

Adm. Ernest King, suspicious of Eberstadt's alliance with the navy secretary, ignored the report as well. When testifying before the SMAC, he reiterated the traditional doctrine of the U.S. Navy as the first line of defense and best means for national security. King focused on what Maxwell Rabb, Forrestal's personal observer at the hearings and a Unification Study Group member, called "the old line Navy approach." King offered nothing new and was "somewhat of a disappointment." Willett, who monitored the

SMAC hearings as well and sent Eberstadt a daily summary, thought that despite a favorable response accorded Forrestal's introduction of the Eberstadt Report a few days earlier, King had refused to follow up on the interest. "He did not tie in closely enough with the Eberstadt program as presented by Forrestal," Willett observed. In fact both former study group members agreed that the navy had presented neither "constructive steps" nor an alternative to what Rabb called the amalgamation of the armed services.[20]

Frustrated by the navy's apparent indifference to his report, Eberstadt worked behind the scenes during the SMAC hearings to promote his plan. A new sharpness marked his treatment of unification, attesting to Eberstadt's growing conviction that his plan transcended advice to the Navy Department that could be dismissed simply as a tactic to offset the War Department's plan. In a draft of remarks that he prepared for Carl Vinson, House Naval Affairs Committee chairman, on October 20, Eberstadt insisted that the single-department form of unification constituted the most insidious and dangerous challenge to the U.S. Constitution in American history. He called it a direct path to a military takeover of the state. When H. B. Swope chided his friend gently about what he saw as the unfounded fear that unification would lead to "a man on horseback" and "erection of a military power that will overwhelm our democracy," Eberstadt shot back that if Swope did not want real national security then he would support the "thin film of unification." Eberstadt lectured the elderly Swope: "I don't happen to think that one department and one secretary, which is as far as the advocates of unification are in agreement, is even an approach let alone a solution to our present problems."[21]

At the same time, Eberstadt promoted his report among the best of the Good Men on Forrestal's staff: H. Struve Hensel and Artemus Gates. They responded by pushing Eberstadt's study as the official Navy Department view. Undersecretary Gates urged Forrestal to incorporate the plan as his own when he appeared before the SMAC. "Instead of referring to the Eberstadt plan," he advised, "I feel you should adopt it in principle and expound it as the Navy plan, and lead off with it." Gates told Forrestal that if the Navy hoped to counteract the popular army plan for unification, it needed a new approach to the "dead hand of tradition" that blindly followed Mahan's doctrine of sea power.[22]

Hensel consulted more closely with Eberstadt. The latter reviewed Hensel's testimony for the SMAC and drafted speeches for the assistant secretary of the navy. Eberstadt's corporatist assumptions permeated Hensel's address to New Jersey realtors at their Atlantic City convention. In it, Hensel defined national security as the balancing of production and transportation, not the unification of military departments. In his two appearances before the SMAC, Hensel revived the Eberstadt Report, first introduced by Forrestal but obscured by the Collins and Eisenhower testimonies. The assistant secretary introduced the report's author as an impartial outside expert on national security who had dedicated six months of intensive study to the very questions that now puzzled the Senate committee. The Navy Department never restricted Eberstadt, Hensel insisted, saying that Forrestal's advisor "was given absolute freedom in expressing his opinion." And that opinion concluded that national security came not from military unification but in the development of the integrated management and decentralized structure of a coordinate system that would link the military, foreign, and economic policies of the state. Hensel suggested that the Collins Plan and the War Department (and by implication the president) lacked basic understanding of these larger national issues and problems. "After I had finished," Hensel wrote Eberstadt, "there were no questions about the Collins Plan and even Senator Hill in his examination of me said, 'Let's forget the Collins Plan!'"[23]

Meanwhile, Eberstadt worked through members of his former study group team, some still on active duty with the Navy Department and updated him daily on the most intimate departmental business connected with reorganization and unification. Like Gates and Hensel, they sought to transform the Eberstadt Report into a navy plan for national security. Scott Harrod informed his study group chief that after consultation with E. F. Willett, Maxwell Rabb, and Eugene Catron, they had decided to draw up a sales brochure that synthesized the report. "I visualize the little pamphlet, nicely printed in rather large primer-like type," Harrod wrote. After Truman ignored Forrestal's request for a Morrow-style commission to study national defense and started work on his own unification proposal, Forrestal ordered mass production of Harrod's brochure. "There is much rushing around and activity here today," Willett confided to Eberstadt on November 30. "I guess the Navy wants to make a splurge before Truman takes a position. It is planning 1000's of copies of a collection of charts and anti-unification propaganda by Monday," he wrote. A week later, the glossy magazine version of the "Navy Plan for National Security," complete with red-white-and-blue charts, appeared in print. Catron prepared one hundred copies for Eberstadt's personal use and promised more.[24]

Eberstadt's continued access to the naval bureaucracy depended upon that intimate personal bond with Forrestal that Wallace Macgregor had noticed at the study group luncheon in September. Forrestal gave him immediate entrée to his office and to the Navy Department any hour, for any reason, though Eberstadt held no official position with either the navy or the government. From time to time Forrestal dispatched a navy plane to fetch his friend and always sent him copies of every confidential document that dealt with unification. Eberstadt's influence increased during late 1945 as Forrestal struggled to weave a course among naval, administration, and corporate organizational interests, while protecting naval integrity and organizing for postwar security. Eberstadt prepared his partner for committee testimony much as he had in 1933 for the Senate Banking Committee investigation of Dillon, Read's financial deals. He briefed Forrestal on how to avoid trick questions from pro-army senators. He recommended names of additional witnesses who could defend the navy's case. Eberstadt explained how to indoctrinate CNO Chester Nimitz, Thomas Robbins, and other admirals for the congressional hearings so that they could withstand cross-examination from hostile senators. He advised Radford, Sherman, and other officers who assisted Forrestal in preparing the navy's position on unification and defense reorganization. Eberstadt drafted a concurrent resolution that called for the establishment of a Morrow-like board to study postwar national defense. He prepared Forrestal's response to accusations that the navy had not consulted with the army on postwar planning. Eberstadt reviewed nearly every departmental statement on unification before release to the press.[25]

Above all, Eberstadt continued to press Forrestal to consider "the possibility of the Navy getting 100% behind the report as a program which in breadth and scope far transcends the narrow issue of unification and which sets forth a program of team-work and unity on the whole national security front." Eberstadt worried that if the secretary delayed adopting the Eberstadt Plan, unification proponents would seize selected parts for their own purpose. Eberstadt asked his friend to study reorganization by establishing a secret section, headed by Hensel, aided by politically astute air admirals Arthur W. Radford and Forrest Sherman, assisted by members of Eberstadt's study group, and supplemented "by a first class publicity man." This organization, inspired by Eberstadt and expanded by Forrestal, became the Secretary's Committee of Research on Reorganization (SCOROR), which played a pivotal role in preparing the navy's position on unification until passage of the National Security Act of 1947. Eberstadt recommended as well that the secretary mobilize the Navy League and Navy Industrial Association, two private interest groups with which Eberstadt had active association, to present "their objections from a local point of view to unification."[26]

Eberstadt became more aggressive in late November, 1945. He wanted to come to Washington with John Cahill, a partner in Eberstadt's old law firm and close friend and legal consultant for Forrestal, to lead the navy's fight against the single department form of unification. He thought the job too big for Hensel and for the admirals. Eberstadt suggested for the first time that unification was a life-and-death situation that required his personal intervention: "I am convinced from the attitude and words of the air people, and perhaps to a lesser extent from the Army, that if unification goes through, our sea power will be dealt a mortal blow." Eberstadt's usually calm and moderate approach, which had steered Forrestal through many little problems and larger crises, became shrill, breeding suspicion and uncertainty in his already embattled friend. Eberstadt's change in tone perhaps reflected the influence of Baruch, whose South Carolina estate Eberstadt had visited, or arose from concern that Forrestal's interest in the Eberstadt Report seemed to be fading. Most likely, it came from the knowledge that the president had started serious work on his own national security plan. Eberstadt suspected that Truman's plan promised to further concentrate central state power and regulation begun in the New Deal, rather than develop the "art and practice of cooperation" between national interests—the main feature of his own plan for national security. Eberstadt's sense of urgency over unification coincided with news that Truman was considering a message that would be the final word on the subject.[27]

Truman compiled a unification message in November, months earlier than originally expected. Though he approached the issue with less emotionalism and sense of crisis than either Forrestal or Eberstadt, Truman showed signs of impatience and concern with what his friend and naval aide, James K. ("Jake") Vardaman, termed the "stubborn attitude" of the Navy Department before the SMAC. Presidential counsel and speechwriter Sam Rosenman warned Truman of the "grave danger" of allowing a Cabinet officer and the navy to continue opposing their commander-in-chief. Moreover, the president had seen a confidential memorandum from H. B. Miller, navy director of public information, to all flag officers that encouraged opposition to unification and the administration's plans for reorganization. Such reports angered Truman. As he told Harold Smith, he felt that the navy had "doubled-crossed" the president.[28]

Truman displayed increased sensitivity to the flood of anti-unification mail that arrived in the White House, some of it stimulated directly by Forrestal and Eberstadt, who had asked people on the Good Man lists throughout the country to write the president and express their opposition to the single-department plan. These letters from private individuals and associations, Catholic leaders, and labor organizers called Truman everything from a cruel, Nazilike dictator to a pawn of the communists for pushing defense unification. Some compared him unfavorably to FDR. "I feel quite certain, that our greatest President, Franklin Delano Roosevelt, would not combine all forces, and tend to make it monopolistic," one message announced. Finally, when Navy League official Sheldon Clark accused Truman of plotting to establish rule by a military clique, the president exploded. "Never read so much bunk in so few lines," he snarled.[29]

Newspaper and radio commentators, particularly those in frequent communication with Eberstadt and Forrestal, harped on the drift in military policy and lack of presidential leadership in national security affairs. Radio analyst Raymond Gram Swing urged his ABC listeners to give the navy plan a fair hearing before backing Truman's unification designs. Lippmann demanded that the president lay down guidelines for unification, so that people like Doolittle could be controlled. Even the pro-unification *New York Times* supported Forrestal's call for a presidential commission headed by Baruch to settle national security squabbles. As the navy received more favorable publicity, its approach to national security organization gained credence. Meanwhile, the image of Truman and his War Department allies slipped. Even General Marshall received criticism in the press for his reported role in bringing on the Pearl Harbor surprise attack. This damaging information had been revealed in the Eberstadt Report, and though Eberstadt assured Marshall personally that his report had never blamed the chief of staff for the attack, the damage had been done. This "turbulent situation" had gotten completely out of hand, Smith told the president.[30]

The budget director urged Truman to take a firm public stand on unification, similar to the views that he had expressed in his *Collier's* article in 1944. Smith warned that the interservice feuding, bad publicity, and disintegrating Democratic Party discipline threatened to undermine executive authority and the office of the president. He knew the source of the problem: "I am beginning to feel now, as we face a relatively tremendous peacetime budget that the Navy should not have been permitted to go up on the Hill—even without commitment—independent of a thorough consideration of our international commitments in relation to the size of the Army, Navy, and Air Forces." Smith thought that unification would solve the entire problem. He pressed the president to introduce a unification message at once in order to establish the principle of strong executive management and centralized administration. Presidential counselor Sam Rosenman agreed and placed "merger of the armed force" at the top of his November agenda of recommended presidential messages, ahead of Social Security, federal aid to education, conservation of electric power, and other urgent national concerns.[31]

Truman's staff assembled material for a unification message in early November, 1945. When White House assistant Elsey returned from Harvard the first of the month, he found a note from Clark Clifford, an assistant naval aide, requesting immediate preparation of a file on unification "for the President's consideration." There was a note of urgency. "This to take priority," Elsey observed. "What a headache." A week later, he submitted a loose-leaf unification file with his own study of foreign military organization, an outline of legislative action on unification, and summations of the Richardson and Eberstadt reports. Armed with this material, and information provided by the War Department and Bureau of the Budget, Rosenman started work on Truman's unification message. He assumed that the president wished to adopt the "Army view," and as Elsey discovered, consulted more with the army and AAF representatives while preparing the message than with representatives from the navy. Forrestal knew that Truman held "long talks" with General Marshall. Even Smith believed that Truman identified too much with "the so-called 'Army Plan,'" and the budget director advised the president to take a more independent stand, since his World War I record had labeled Truman as an army advocate.[32]

Clifford and Elsey considered it unwise for the president's unification message to adopt the army view. Both advised Truman to weigh all information and positions without taking sides. They worried that if the president represented the army's single-department position, defense reorganization would surely fail. So in late 1945 both White House assistants advanced the Eberstadt Report and the Navy Plan for National Security because they saw in the larger national program an opportunity to help the president gain public support and passage of a national security law. They never chose sides. "Politics don't mean a thing to me," Clifford claimed, while Elsey saw himself as nothing more than a White House research assistant. "I did a great deal of fact finding and rummaging around for Mr. Clifford in this area," he recalled. Nevertheless, by promoting the Eberstadt Plan against the solution favored by other close Truman advisors, they created a discernible schism in the White House over the form and substance of national security organization and unification.[33]

Clifford and Elsey easily sympathized with the Eberstadt Report and navy plan. A Saint Louis corporate and labor lawyer, Clifford had served as a naval supply officer and assistant to the commander, Western Sea Frontier. Elsey graduated from Princeton, pursued a graduate degree in history at Harvard, and joined the navy in 1941. If their wartime duties had placed them in Forrestal's office rather than the White House, where Clifford served as assistant naval aide and attaché and Elsey as a watch officer in the White House map room, these naval reservists would have fit the mold of the Good Men recruited to the Eberstadt study group. Indeed, Elsey wondered why in October Forrestal had not pressed the Eberstadt Report more forcefully on the White House. "I think this an astounding strategic blunder, because Secretary of the Navy should ensure that he gets support for naval views from White House before army sells its ideas here," Elsey commented. He immersed himself in the Eberstadt Report, discussed it with people in the Navy Department, and listened to Admiral Leahy's views. "Ad. Leahy says merger is solely due to army's anger at not getting money between the wars and they want to be sure to get it from now on," Elsey noted after one of his conversations with the influential admiral. In the end, Elsey advocated the Eberstadt Report. "I hope that this statement will be given very careful thought by the Commodore [Vardaman] and by Judge Rosenman," he told Clifford. "I also hope that it reaches the President."[34]

The two naval reserve officers on the White House staff advanced Eberstadt's report more actively as the unification message neared completion. A week before Truman actually presented the unification announcement, Elsey prepared (and Clifford endorsed) a memorandum that listed a number of their disagreements with Rosenman's draft. The Elsey-Clifford note employed the language of the Eberstadt Report: "If the President takes the approach that unification is necessary because this government would then be in a better position to integrate its foreign, military, and economic policies and that this integration is our greatest hope against our becoming involved in another war he will be making a position, [and a] constructive and optimistic statement well-designed to win Congressional and public support." Elsey suggested that the presidential message treat national security more broadly, that it should look beyond the simple merger of the War and Navy departments. He disagreed with Rosenman's army-influenced view that the entire problem of national security stemmed from interservice disputes. Elsey explained instead Eberstadt's thesis that national security lay in proper planning for industrial and human mobilization. In sum, the Elsey-Clifford paper urged that the Rosenman draft include Eberstadt's recommendations for a National Security Resources Board, National Security Council, CIA, and permanent JCS.[35]

Apparently, Rosenman brought the Elsey-Clifford memorandum to the president, who endorsed the ideas of a CIA and National Security Council. Truman was already committed to a central intelligence group and probably adopted the NSC concept to appease Forrestal. Rosenman seemed uncertain, though, how best to combine these coordinating elements from the Elsey-Clifford digest of the Eberstadt Report with the essence of the War Department blueprint favored by Truman. He had struggled with such organizational compromises for FDR during the war, particularly when he designed the Supply Priorities and Allocations Board. "No man even to the slightest degree contaminated by organizational experience could have formulated, much less advocated anything like SPAB," Eberstadt observed.[36]

This time Rosenman sought the advice of the "neo-Hamiltonian" administrative specialists from the Public Administration Clearing House, the University of Chicago, and similar agencies that had entered the Bureau of the Budget with Harold Smith. Arnold Miles, Leonard Hoelscher, James L. Sundquist, and other experts had assisted Marshall in War Department reorganization and promoted the movement toward executive centralization through the single-department model. Moreover, they had been warned earlier about Eberstadt's alleged "lust for power." Naturally, Miles criticized Eberstadt's design. It was unsound, poorly conceived, and a threat to presidential authority, Miles explained. Eberstadt's structures defied what administrative experts considered time-tested theory, drained legitimate functions from central institutions, and endangered the democratic process by fostering elitist military and industrial power groups outside of the normal organizational channels and lines of authority. The budget official urged the White House not to endorse "the specific provisions of the Navy Plan."[37]

Truman's "Special Message to the Congress Recommending the Establishment of a Department of National Defense" on December 19, 1945, reflected the budget bureau's influence. Sundquist claimed that bureau recommendations formed the basis for the president's entire plan. Truman recommended in his message the combination of War and Navy departments into "one single Department of National Defense" responsible for the "armed national security." He proposed three military branches, for air, land, and sea, with the navy keeping its own aviation and the marines but being merged into the single department under a secretary of national defense, who would replace the secretaries of war and navy in the Cabinet. Truman advocated a single military chief of staff and three commanders for the "component branches." He asked for authority to further centralize, when necessary, military and civilian functions of national defense. The president called for a great "federal research agency" tied to the single department.[38]

Truman's extensive experience with army organization, his New Deal statist philosophy, and his Senate experience with military affairs and investigation of procurement permeated the message. Truman agreed fully with Marshall that this unified defense organization eliminated interservice conflict, duplication, and overlapping of functions. It ensured unified military policy, budget, and supply and an "integrated strategic plan." A single department provided the ideal vehicle to implement UMT for a "well-trained citizenry," a program that Truman had already introduced to Congress in late October, 1945.[39]

Truman presented his message as the administration's final position on national defense reorganization. The subject had been exhaustively debated in the JSC special committee study, the SMAC hearings, the Eberstadt study, and the newspaper press and public forums nationwide. Truman had adjusted his views, incorporating some of the suggestions made during the unification debate, particularly in his assurance of the continued integrity of naval aviation and the U.S. Marine Corps. He had borrowed from the Eberstadt Report when he mentioned in his message the interrelationship of military reorganization with foreign policy, industrial mobilization, and scientific research. Now the debate was over. "Further studies of the general problem would serve no useful purpose," Truman insisted. He expected everyone's cooperation no matter what their prior views were on defense reorganization: "However strong the opposition that has been expressed by some of our outstanding senior officers and civilians, I can assure the Congress that once unification has been determined upon as the policy of this nation, there is no officer or civilian in any service who will not contribute his utmost to make the unification a success."[40]

It was not this simple. Truman's message revealed substantive philosophical and organizational differences with the concepts presented by the Eberstadt Plan. Instead of clarifying issues and ending conflict, his statement on the single department of defense sharpened once again the ongoing political-economic and national security debate that raged during the early stages of mobilization and during the war and that intensified with postwar planning. Although this debate became blurred and intertwined in purely military questions, the defense reorganization fight revealed that in many ways the divisions over the Truman and Eberstadt unification plans were an extension of the struggle between New Dealers and dollar-a-year corporatists over mobilization forms and policies, wartime production organization, and reconversion timetables.

The Eberstadt Plan advanced collaboration, coordination, and consultation among government agencies, between government and business, and between the military and industry through clusters of formal and informal coordinating structures–what Eberstadt called the "ligaments of coordination." Truman's plan mentioned none of these bodies. He introduced no boards for resources, joint research and development, or education and training. Perhaps aware that coordinate forms opened opportunities and became platforms for those dollar-a-year types that he had so "lustily" despised during the war, Truman eliminated such structures in his message. Instead of a joint research and development board with three assistant secretaries to develop the broadest possible contacts between government, private industry, and academic institutions, as proposed by Eberstadt, Truman recommended one great government research agency welded to the single department of national defense. As to the other structures, Truman wanted the number reduced through consolidation and simplification. "They are simplified as the number of these agencies can be reduced," Truman told Congress.[41]

Truman's elimination from his plan of the Joint Chiefs of Staff–an agency that Eberstadt and Forrestal wanted made permanent through legislation–underscored the philosophical and organizational gap between the two approaches to national security. In the Eberstadt Plan, the JCS became that corporate body where representatives of each service could meet, negotiate, and iron out a consensus on national strategy. "In other words," Forrestal added, "the Joint Chiefs of Staff, while it was accused of being a debating society, for that very reason probably was the most practical and useful device that could have been conceived for the conduct of this war." But to Truman, it was "a committee which must depend for its success upon the voluntary cooperation of its member agencies." Committees for "voluntary cooperation" and association, Truman insisted, were "obsolete organizational forms." Clear lines of authority between military and civilian embodied in his chief-of-staff design better fit Truman's conception of the presidency and governmental organization.[42]

The Truman and Eberstadt plans raised different conceptions of democratic politics and institutions. Both recognized that postwar America had to assume greater international responsibilities for the peace and security of the world and that the American public had accepted this role. Both expected the state to play a central role in that process. Here the similarity ended. Eberstadt's plan presented a pessimistic perspective on how public opinion, press, and congressional politics would respond to the idea of "implementing our ideals for world order with the use of force against aggressor states." Left to themselves, pluralistic interest groups feuded, split into factions, and fragmented national policy. Eberstadt thought that under such conditions it became impossible to develop a consistent and coherent national policy. In fact, his plan bypassed democratic institutions. Instead of involving the public (or elected representatives in Congress), Eberstadt's plan appealed to "groups particularly concerned with aspects of military policy" and provided machinery to educate, inform, and inculcate a "full and correct understanding" of national security. The NSC, NSRB, Education and Training Board, and even the CIA had educational and consensus-building functions.[43]

Truman's plan expressed faith in the democratic process and in the American people to understand and accept the "continuing burden of responsibility for world leadership." In his scheme, the American public, through its elected representatives in the White House and Congress and through its responsible political parties, press, and interest groups, would support a unified national security policy. The single defense department format allowed these elected officials to bring the full force of government to bear on national welfare and security. The American people trusted the president, Cabinet, and Congress to formulate and conduct military and foreign policies because, as Truman argued, they understood that these reflected the "wisdom on all subjects that affect the political life of the country."[44]

Truman offered his plan as the best path to civilian control over the military through a single channel of authority. His national defense blueprint retained civilian control in one central office rather than dispersing it, as in the Eberstadt Plan, through many amorphous corporate coordinating agencies. Truman implied that special interests and militarism would lurk in such agencies, uncontrolled by elected, responsible government. Directly answering Eberstadt's charge that a single secretary would become a puppet of the military establishment and undermine the constitutional authority of the president, Truman argued that the American public had the "solid good sense" not to permit such a military dictatorship.[45]

Designed to protect democratic institutions and to quiet controversy, Truman's unification message stimulated conflict instead. Rosenman's final product suffered from the same organizational weaknesses manifested in his supply board design for FDR in 1940. By incorporating concepts from the Bureau of the Budget, the Eberstadt Report, AAF, the War Department plan, and Truman's *Collier's* article, the message satisfied no one and disappointed almost everyone. The War Department felt that the report had not gone far enough, and the budget office disparaged compromise with the Navy plan. The AAF supported the president's message only if it secured an independent air force. Robert Hannegan, postmaster general, thought it politically damaging, and congressional critics vowed to fight it. Elsey and Clifford seemed deeply disappointed. They had received "small crumbs" for all their work and advice, and Clifford worried that Truman had made the Navy Department "desperately unhappy." Indeed, Admiral Leahy insisted that the navy would never accept the organization suggested in the president's message.[46]

Eberstadt reacted almost violently to Truman's unification message. He called the president a fool for throwing away the opportunity to unite feuding factions. He had lost the chance to become a great postwar statesman and leader. "The difference between a fool and a wise man," Eberstadt wrote Forrestal on December 20, "is that the latter foresees the reasonable consequences of an act or situation and the former does not suspect what is going to happen until it actually occurs." Truman displayed the former characteristic. Eberstadt believed that the president, who was without the typical "Good Man" corporate background, education, or experience, utterly lacked understanding of the need to settle disputes through conference and to seek public consensus. Instead of directing Forrestal, Patterson, and the armed services chiefs to confer like the representatives of great corporate interests in search of a common national goal, Truman had encouraged individual interests to compete, fight, and turn to political intriguing in order to obtain power.[47]

Sensing that Truman's statement had foreclosed his own plan for postwar national security, Eberstadt reminded Forrestal that they had to do their utmost to avert "the serious consequences of unification." Eberstadt did not want to follow the lead of his elderly friend Baruch, who had seen his vision of corporatist institutions fade after the first world war with the abandonment of the War Industries Board and neglect of industrial mobilization planning and organizations as national crisis and commitment had disappeared. "The lessons of World War I, so fully and accurately recorded by

Bernard M. Baruch," Eberstadt wrote, "were at the outset practically completely disregarded."[48]

Eberstadt would not let this happen again. He would do whatever he could to avert a similar national disaster this time. He would mobilize Good Men from labor, industry, business, and religion to fight unification. He would contact his friend George Meany, Stock Exchange boss John Coleman, and Catholic religious leaders in New York. He would collaborate with "our statesman friend, Carl Vinson," and with new members of the Senate Naval Affairs Committee, H. Alexander Smith and retired admiral Thomas Hart. He would never allow Truman's message to discourage Forrestal's commitment to the Eberstadt Plan. Eberstadt appealed again to his best friend to continue the fight against unification and against the president's plans for reorganization. "We have been through a couple of tough spots together and now we are in another," he confided. "I made the statement to you that you had the better case and couldn't lose."[49]

Eberstadt's unusual passion disconcerted Forrestal. "He is anxious to get into this debate with both feet," the navy secretary told former assistant Ralph Bard, now the president of the Navy League. "I don't feel that I'm free to encourage him in that, so I suggested that the Navy League was the proper and obvious vehicle through which he should do whatever he's going to do," Forrestal said. His position in the Truman administration limited his ability and desire to oppose policy. He thought the navy could no longer advance nor discuss the Eberstadt Plan in public. "I don't care, of course, whether I stay in this place or not another day longer," he told Bard, "but I don't want to be fancy with the President."[50]

Truman's unification message caught Forrestal between the president and his best friend, two strong personalities to whom he owed, respectively, official obligation and personal, Good Man loyalty. While Eberstadt pressed Forrestal to fight his boss, Truman worked on Forrestal's intense patriotism, personal ambition, and loyalty to the administration. Truman invited the navy secretary to the White House for little chats and pep talks, placing the burden for postwar national security squarely on Forrestal. The president told him that his support for unification was vital. "He said it might mean the future peace of the world or an armament race," Forrestal noted in his diary. Truman lectured Forrestal on how unification provided the most democratic, efficient, and economical solution to the entire problem of postwar national defense. Modern war required the combination of the armed services into a single establishment; according to his diary, he wanted "to wrap the entire question up into one package and present it to Congress." General Marshall supported the president completely, Truman explained, and the president wanted his navy secretary to show the same loyalty.[51]

Truman appealed to Forrestal again just before the public release of the unification message. He reminded Forrestal that he had allowed the navy to debate unification openly and to express its views fully. Truman had personally studied the Navy plan and had considered it in his message by ensuring the continuation of naval aviation and the Marine Corps organizations. Truman insisted that his single-department plan provided for better coordination and security than that presented by the navy but that he had no intention of threatening the navy's position in the defense establishment. "I believe we can work out a program that will in no way handicap any of the services, and will still make it possible for the President to be in complete communication with the whole set-up when it is necessary," Truman promised.[52]

Truman had Forrestal convinced for the moment. The secretary considered the president's approach to be "entirely fair" and based on an honest conviction that consolidation offered the soundest approach to national security. After such exhortation, Forrestal showed a renewed willingness to go along with the president. He refused in late December to encourage Eberstadt or the Navy Department to pursue anti-unification activities. When one advisor gave Forrestal a statement to clear with the White House, a statement suggesting that naval district officers be allowed to express views contrary to the president's message, Forrestal wavered. "I would be inclined not to send this along to the president–if I were in his place I think I might be a little bit irritated by such direct evidence of returning to the attack," he wrote. Forrestal encouraged the promotion of sea power and amphibious doctrine, however. He thought that the president would not mind a "few shots from the standpoint of politics."[53]

Forrestal's resolve to curb anti-unification activity and to stop promoting Eberstadt's plan was short-lived. When he returned from a brief Florida vacation in early January, 1946, Forrestal resumed the campaign against Truman's plan for national defense reorganization, mostly through indirect channels such as Eberstadt, who now became most active in the politics of unification. Eberstadt encouraged his friend to resume the battle. "James, remember when I said to you when they really go into this you have no chance of losing," he told the secretary as soon as he had returned from Florida. Forrestal needed little prompting. He believed that the army had forced the president to "let go of the Message at the time he did because they realized we were making more headway than they liked." Worse still, by rushing Truman to make a premature unification statement, the army had secured official endorsement for its narrow view that national security could be obtained through centralization. The simple fact of U.S. responsibility for global stability justified in Forrestal's view the renewed agitation against unification. Forrestal determined to set the president on the right course toward a national security system that linked all elements of the state's power.[54]

While on vacation (his first in nearly five years), Forrestal had brooded about the implications that unification seemed to hold for the development of postwar national policies. "One cannot help but be struck by the tremendous task that is involved in the implementation of American policy," he wrote on January 2. "They are the difficulties that always confront the liberal and reasonable mind in attempting to construct a workable pattern for international relationships in the complex of racial, religious, and vastly different social-political traditions and habits," Forrestal confided in his diary. Truman's message offered no structure, blueprint, or solution for this complex task or indeed with which to confront what Forrestal saw as the emerging struggle against Soviet Marxism. His ruminations reflected recent correspondence with Walter Lippmann and W. A. Harriman about communism as well as an essay on dialectical materialism prepared for him by Eberstadt's assistant, E. F. Willett. "Our task is further complicated," Forrestal wrote, "by the fact that we are trying to preserve a world in which a capitalistic-democratic method can continue, whereas if the Russian adherence to truly Marxian dialectics continues, their interest lies in a collapse of this system."[55]

Forrestal placed the form and structure of national security in this political-economic and ideological setting of world confrontation. The Eberstadt Plan provided the tools for those who, like Forrestal, were "trying to preserve a world in which a capitalistic-democratic method can continue" as well as the machinery with which to resist and contain what Forrestal called the "locust-like effects" of "the Russian colossus." The army plan for unification, on the other hand, offered only one structure, one channel, and one option with which to meet complex national and international obligations and threats. It seemed to Forrestal, as well, that those who pushed unification were "out to wreck the military power of this country" by destroying the navy, accelerating demobilization, and sharing atomic secrets with the Soviet Union.[56]

By reducing American ability to mobilize its national power, friends of unification tolerated worldwide disorder, malnutrition, and unemployment, Forrestal argued. "A lot of the support for [unification] comes from very violently left wing people," he confided to Lewis Douglas. "I don't say that they have anything in mind but we know that there are some very curious people in the Army, to say the least," he added. Those curious people, perhaps unwittingly, wanted a plan that undermined national strength at home and abroad. "And if you get this big global amorphous mass," Forrestal concluded, "it lends itself that much more readily to infiltration."[57]

Forrestal had to resume the fight for the Eberstadt Plan as the alternative to "this big global amorphous mass" of unification that threatened to swallow the navy and the nation. Eberstadt's coordinating mechanisms afforded an organizational model that allowed the navy to defend in the future (as it always had in the past) Americans' freedom to trade and travel through control of the seas and air space over the oceans. Eberstadt's system preserved naval autonomy so that it could protect and promote American foreign, military, and economic programs for international stability and world peace. In Forrestal's view, the navy's global capabilities became more urgent than ever in the emerging struggle to stop Soviet expansion. The Eberstadt Report introduced the necessary machinery for this struggle, and over the next eighteen months the building of that machinery became a central concern to Forrestal. Little else had so obsessed him as this unification question. At every turn during this period, Forrestal drew upon Eberstadt's advice, help, and organizational philosophy in the drafting of America's first national security law.

8. Shaping America's First National Security Law, 1946–1947

Eberstadt and Forrestal shared a moment of personal triumph. Despite Truman's unification message, they had slowed the army's unification blitz by convincing the navy, press, and Congress to consider the Eberstadt Report as an alternative to the War Department's single-department program. They had placed postwar national security in the larger political-economic context of mobilization, technological and scientific developments, and corporate business-management principles. Eberstadt praised Forrestal for orchestrating a successful publicity campaign that promoted this broader national security view and deflected the army scheme. "It has been outstanding and has, in my opinion, been the most effective flank attack against unification," he assured an obviously pleased Forrestal, who was always buoyed by his friend's confidence in him.[1]

Forrestal appeared to be at the peak of his influence, power, and administrative skill. The president found him indispensable. "I thought too highly of him as a public servant to allow him to resign," Truman insisted. The navy secretary had generated tremendous public recognition during the war and by 1946 had become a leading national figure portrayed as a tough, honest, and charismatic official. Photos showed the ruggedly handsome Forrestal immaculately dressed and groomed, usually with a pipe clenched firmly in his teeth, a stark contrast to the somewhat tired and sad-looking Truman, who stared out through thick glasses. Forrestal's popularity, close friend William Douglas assured him, meant that he could become New York governor, senator, or take any job that he wanted. Some mentioned the navy secretary as a possible presidential candidate. No wonder that on January 7, Forrestal found himself in "a smug mental condition—today."[2]

He had reason for this upbeat and optimistic attitude. Negotiations between CNO Chester Nimitz and Eisenhower, the new army chief of staff, promised to settle out of court what Forrestal called the unification lawsuit. "Eisenhower is a good practical Dutchman and so is Nimitz, and between them I believe we will make progress," Forrestal wrote. Probably unaware that Eisenhower espoused a form of the single-department doctrine that relegated the Marine Corps to the role of a tiny police force, Forrestal found the general warm and charming and much easier to work with than his aloof predecessor. Forrestal admired Marshall but felt intimidated by him. Eisenhower sat down with the navy to iron out joint procurement, intelligence, research and development, training, and education programs, seemingly without reference to the odious single-department format that had dominated Marshall's approach. Eisenhower "couldn't have been more responsive to the idea of getting to work right now," Forrestal told Lew Douglas.[3]

The progress of Secretary's Committee of Research on Reorganization pleased Forrestal as well. Originally suggested in discussions with Eberstadt during the autumn, SCOROR had become a full-fledged bureaucratic machine with sixty-eight permanent employees divided among functional sections and commanded by four flag-rank officers. SCOROR staff included naval public relations specialist Walter Karig, USMC representative E. Hunter Hurst, and five Eberstadt study group members. The latter linked Eberstadt to this highly confidential naval organization and provided copies of secret SCOROR documents. They sought his advice on the group's mission of "indoctrinating the naval service, as appropriate, with historical, legislative, economic, and other facts concerning unification of the armed forces, and the problems incident thereto." Using Eberstadt's report as the source of information, SCOROR flooded the country with navy organizational doctrine. The Third Naval District commandment held forty-seven luncheons in New York City to present the Eberstadt Plan. The Ninth Naval District distributed a folder containing material from the Eberstadt Report to officers as they returned to civilian life. Massachusetts congressman John McCormack told Forrestal that the grass-roots campaign was working, since "there isn't any talk up here about your unification which is a good sign."[4]

SCOROR fed material to Eberstadt and Congressman Harry Sheppard of California, who was chairman of the House Naval Appropriations Subcommittee, for a debate on the national radio show "American Forum of the Air." Eberstadt and Sheppard were to debate against Jimmy Doolittle and Sen. Edwin Johnson of Colorado, an SMAC member and single-department advocate. Thoroughly briefed by SCOROR, Eberstadt demolished Doolittle's arguments in support of a single department of defense by employing data that revealed the War Department's own internal organizational weaknesses. After the radio debate, one senior admiral applauded Eberstadt's brilliant presentation, calling it a masterful victory for the navy.[5]

Increasingly, Eberstadt assumed a powerful behind-the-scenes role in unification politics. He appeared during 1946 at every intersection of power and decisionmaking in the unification controversy and in the creation of a national security structure. While Baruch once dreamed of such influence, Eberstadt achieved a large measure of it as an organizational facilitator between bureaucratic sectors and interest groups directly involved with immediate postwar national security reorganization. He served as the intermediary agent between Forrestal and Robert Patterson. He arbitrated meetings attended by navy, army, and AAF representatives. Eberstadt counseled legislators, wrote drafts for a national security law that included the corporatist structures outlined in his report, and prepared the SMAC minority opinion that opposed the single-department plan recommended by the committee's majority. He advised admirals, settled differences between naval aviators and more traditionalist line officers, and guided the naval establishment toward the Eberstadt Plan. He prepared Forrestal's civilian staff members for bureaucratic battle, though his own study group experts had returned to civilian life and a new set of assistants had arrived in the secretary's office. Unrestrained by loyalty to the Truman administration, Eberstadt spoke against unification freely—on the radio, on the podium, and through frequent and effective correspondence with the nation's leading journalists. In many subtle and direct ways, Eberstadt left his mark on the first American national security law by creating conditions whereby the president had to incorporate more of Eberstadt's corporatist structures if he hoped to secure Forrestal's support and gain national defense reorganization in 1947.

Eberstadt's intimate association with Forrestal gave him access to the bureaucracies, accelerated involvement in administrative politics, and determined the direction of his contributions. Instead of remaining a "kibitzer" like his elderly mentor Baruch, Eberstadt entered the doors opened by Forrestal to affect policy and structure. After Forrestal's death in 1949, Eberstadt's inability to cultivate similar channels through Eisenhower attested to the importance of the Forrestal-Eberstadt connection. In January, 1946, Forrestal opened another avenue when he arranged for Eberstadt to assist the SMAC subcommittee of Senators Elbert Thomas, Lister Hill, and Warren Austin with the drafting of a unification law. Austin welcomed Eberstadt's help. The internationalist Vermont Republican, later the first U.S. ambassador to the United Nations, memorized portions of the Eberstadt Report and considered its arrangements for linking foreign and military policies an essential framework for compromise in any national security legislation. He assured Eberstadt that they would write the law from scratch without prior commitments to any interest. Austin explained that they would base legislation upon the SMAC hearings and Eberstadt's "own admirable report as part of that testimony."[6]

Eberstadt accepted the subcommittee as his "client." The job fit his corporatist assumption about congressional politics developed over the years, from Carl Albert Fetter's lectures at Princeton to the most recent discussions with Herring. Eberstadt considered congressional hearings to be little more than platforms for lobbying and political dealing, both of which resulted in stalemate and conflict. He thought that real progress occurred in congressional subcommittees when a few select legislators negotiated with a few select experts, such as representatives from the Good Man List, to develop programs. Then, these select few hammered out a consensus, educated the public, and secured passage of proposals as a product of democratic decisionmaking. Forrestal favored this type of exclusive committee process, too. Already he had attached Adm. Arthur W. Radford, the SCOROR officer "willing to play the angles," as navy consultant to the subcommittee. With Eberstadt as a private consultant, Forrestal saw the opportunity to persuade the senators that the navy accepted unification, and then to replace the single-department framework with the "over-all Navy Plan for National Security."[7]

Having his people on the inside of SMAC contributed to that smug mental condition that Forrestal had expressed in January. If the SMAC failed to write a national security law, as Forrestal thought it might, such failure resided in congressional politics and not with navy obstruction. This in turn cleared the way for the Senate Naval Affairs Committee (where Forrestal exerted considerable influence) to offer a legislative proposal of its own that more closely reflected the Eberstadt Plan. As the maneuvering continued, Forrestal gradually slipped into depression. His mood plummeted in early 1946 as he took personally the deep "gloom" that descended upon Washington. Commerce secretary Henry Wallace framed the general atmosphere in ideological terms, observing that anyone who held leftist sentiments was labeled a "commie," while anyone with rightist leanings was called a fascist. David Lilienthal saw the pessimism as part of an emerging Cold War and a future governed by atomic weapons. It had arisen "from the terrible realities of the bomb" and "the constant fear of a long, bitter period of antagonism and strain and perhaps war with the Russians," Lilienthal observed in March, 1946.[8]

"Will" Douglas added to Forrestal's growing depression with the observation that Stalin's recent speech announcing a five-year industrial and rearmament plan constituted "the declaration of World War III." Other Forrestal contacts expressed similar concerns during early 1946. William Bullitt, George Kennan, W. Averell Harriman, and Admiral Leahy predicted that if left unchecked Soviet communism would overrun Greece, Turkey, Italy, France, and China. Forrestal's conversations with Winston Churchill a few days after the former British prime minister delivered his "Iron Curtain" address in Fulton, Missouri, plunged him more deeply into a pessimistic state of mind. Forrestal noted that Churchill "was very gloomy about coming to accommodation with Russia unless and until it became clear to the Russians that they would be met with force if they continued their expansion."[9]

From Forrestal's perspective, military strength and a coordinated national security organization had become even more vital for the defense of American interests against this worldwide communist threat and for containing Soviet expansionism. These elements depended upon the eventual outcome of the unification fight, which after a promising beginning, for Forrestal, in January had started to unravel. The Senate subcommittee rejected Eberstadt's advice in favor of that made by Generals J. Lawton Collins and Air Force planner Lauris Norstad. Forrestal cabled Eberstadt, through the U.S. naval attaché's office in Mexico City, where Eberstadt vacationed, informing him that the subcommittee had abandoned their plan. SCOROR would try to recover from the damage, Forrestal continued, but with H. Struve Hensel, Eugene Catron, and most study group members leaving the Navy Department at the end of the month, only Adm. Radford and inexperienced new assistants remained to help Forrestal battle for their organizational plan. The secretary wanted Eberstadt to return to Washington.[10]

Delayed by a family member's illness, Eberstadt did not return to Washington until early March, when he found the Senate subcommittee vacillating between one version of a draft bill that reflected his outline and another that included the War Department format. Efforts to redirect the three legislators to an earlier Eberstadt draft failed. "My interview with the Senate subcommittee and its military associates was disappointing and, I think, almost wholly ineffective so far as influencing the Senators was concerned," Eberstadt said. Rebuffed, Eberstadt urged Forrestal to fight against the subcommittee's final draft. Eberstadt reminded Forrestal that the world confronted an awful situation and that now was not the time to accept the dangerous (and ineffective) organization embodied in the single-department scheme. He exhorted Forrestal to battle "to the last ditch" against any law that included a single chief of staff. He suggested that they draft alternative legislation through the Senate and House Naval Affairs Committees and their statesmen allies (David Walsh, Carl Vinson, and Leverett Saltonstall). Thoroughly agitated with SMAC subcommittee chairman Elbert Thomas, Eberstadt called the navy secretary several times to suggest that Forrestal put pressure on the quiet, scholarly Utah Democrat, who was once investigated by the FBI for alleged leftist leanings. "I think that Thomas is very timid," Eberstadt explained. "I would give it to him strong and unequivocally."[11]

Eberstadt's frustration with the subcommittee mired Forrestal more deeply in that gloom that Lilienthal and others observed swallowing the government. "I think, Eber," Forrestal told his friend over the telephone, "these events of the past six weeks and the condition of the world–to project this subject now just makes utterly no sense." In any case, the secretary called Thomas, as Eberstadt had advised, and lectured the senator that the arrangements written into the subcommittee bill were unworkable. Forrestal then told the president that the Thomas unification bill filled the navy with deep misgivings. According to the navy secretary, the bill created a supersecretary and stuck three "so-called" service secretaries off in an annex with neither responsibility nor power. Forrestal derived his own conception of a secretary of common defense partly from Churchill's recommendation for the creation of an American "Minister of Defense." Such a minister, Forrestal thought, required only coordinating functions and a staff of competent people. He wanted the three service secretaries to retain strong control over their own departments, thus creating something akin to a corporate board of directors in the decentralized operation of General Motors and other great corporations. "From my own experience in business I know that mistakes of judgment are far less likely to occur if the proponent of any plan or idea has to justify his case before a group of intelligent partners," Forrestal said. He doubted that Truman fully understood what he meant: "The President did not seem to draw this line as clearly as it was in my mind, but I think at the end he began to grasp what I was talking about."[12]

During his mid-March meeting with Truman, Forrestal complained about AAF lobbying. The navy secretary, whose SCOROR machine and personal friends and allies conducted their own vigorous campaign against the single department, urged the president to halt such activity. He assured Truman that the navy would stop but doubted whether the War Department could control its undisciplined air arm. However, Forrestal felt that his own promotion of sea power as first defense constituted neither lobbying nor defiance of the president's reorganization policy. "It is obvious that as a member of the Administration, I cannot permit opposition to the announced program of the President, but I do feel that the Navy is entitled to argue the importance of the retention of sea power and of the identity and traditions of the Service," Forrestal wrote.[13]

The air power issue assumed a central part in Forrestal's activities. He recognized a shift during the war from the navy to the air force as the first line of national defense. Forrestal found increased public recognition and favor everywhere for the idea of air power, strategic bombing doctrine, and an autonomous U.S. Air Force. This shift threatened existing structures, military and naval budgeting, and the navy's share of national resources. Forrestal sought to slow this change through institutionalization of the bodies recommended by the Morrow Commission in 1925, including the separate departments with their own integrated air forces, and advocated vigorous government-industry collaboration on research and development of both commercial and military aviation. This concern was, in part, what had prompted Forrestal to send his personal copy of the Morrow Report to Truman a few days after FDR's death, but the new president rejected the idea of another Morrow-type presidential advisory board to make recommendations on national defense reorganization.[14]

Truman's continued neglect of the Morrow commission-of-experts solution disturbed Forrestal. He felt that the president lacked a basic understanding of a balanced defense force; the relationship of sea power to national economic, foreign, and military policy; or the concept of sea–air power doctrine. To address the latter problem, the navy secretary invited Truman to visit the large new attack carrier, the USS *Franklin D. Roosevelt,* for a demonstration. Still, Forrestal believed that the president shared the popular emphasis on air power and favored a strategic air force as the most economical and efficient path to national security in the postwar world. He found Truman particularly susceptible to AAF influences that by early 1946 had, in Forrestal's words, become an "all-out drive" to sell air power doctrine to the nation. "The Air people have been aggressive, ruthless, and somewhat (in my opinion) indifferent to the broader interests of the national security," he wrote.[15]

Concentration on an AAF threat to the navy influenced Forrestal's approach to unification. In his mind, that threat made the unification controversy a national crisis. It even overshadowed the corporate organizational philosophy that underlay the Eberstadt scheme. Forrestal knew that since the beginning of the year the AAF had accelerated dramatically its campaign for autonomy and framed that campaign in the single-department format and around Truman's unification plan. Army airmen placed their struggle in an increasingly hostile anti-navy context. "For the next twenty years, the Navy will be the mortal enemy of the AAF and, indirectly, of national security," warned Col. W. Barton Leach, an AAF planner and Harvard law professor. Air Board Secretary Hugh Knerr added that the navy sought nothing less than destruction of the AAF so that it could dominate postwar national defense. "The amount of money now available to both Army and Navy for these purposes cannot possibly be spent without saturating the resources of the country in both brains and materials, leaving little to the less aggressive Air Force," Knerr advised Carl ("Tooey") Spaatz, the new air force chief of staff.[16]

Spaatz led the fight for his service. In February, 1946, he initiated "a plan whereby Air Force speakers, both on the high and low levels, will try to reach every community of the U.S. to present our side of the story." He began the assault with a talk before the Aviation Writers Association Dinner in which he presented the emerging AAF arguments for unification. According to this view, the world war vindicated a twenty-five-year crusade for a separate air force that was begun by Col. Billy Mitchell, who had martyred his career through court-martial for the cause of an independent air service. Not only had air power won the second world war, Spaatz claimed, it would also deter future war. AAF leaders argued that one hundred heavy bombers deployed on the Philippine Islands in 1941 would have prevented the Japanese attack on Pearl Harbor. Now they insisted that atomic bombs made other services obsolete. Traditional armies and navies lay helpless against the power of modern bombers of V-2 type rockets armed with atomic warheads. Moreover, the United States no longer confronted enemies with strong navies, so a large U.S. Navy was an unnecessary expense.[17]

Spaatz established a group of select officers with public speaking skills to appear before clubs, veterans organizations, chambers of commerce, and national conventions. AAF officers published articles on air power in professional and popular journals, with Spaatz contributing an essay entitled "Air Power in the Atomic Age." The commanding air general wanted at least one item each day on the air force to appear in the national press and ordered the Air Technical Service Command at Wright Field to provide illustrative material to local commands in support of the "president's plan." Radio programs helped, and the AAF Logistic Planning Division prepared scripts for a broadcast that explained national strength, America's world greatness, and insurance against future war in terms of unification and an independent air force. The AAF employed a stable of Hollywood stars, directors, and producers, including Merian C. Cooper, John McCrary, Jr., Frank Capra, William Wyler, Clark Gable, and Jimmy Stewart. In early February, Jack Warner of Warner Brothers Studios sent Spaatz a copy of the film *Unification of Command*. He would show the film at once, Spaatz assured Warner, saying it would be "a great contribution to our effort in putting across this legislation for the defense of the U.S."[18]

A publicity master in his own right, Forrestal recognized that promotion of the navy paled by comparison. This prompted his call in March, 1946, for Truman to turn off the escalating propaganda war. The president sympathized with the AAF campaign, however, encouraging the Air Power League and lobbyist William Redding, while simultaneously condemning and curbing Navy League propaganda. Thus Spaatz increased the intensity of the AAF public relations program. Robert Lovett, the former assistant secretary of war for air, assured Spaatz that Truman supported his efforts. Apparently Ambassador Harriman discussed with Lovett a conversation with the president where they had linked a powerful, autonomous air force to American policy toward the Soviet Union. Harriman explained that while the Soviets did not fear American land or sea power, they were "keenly aware of the power of an adequate Air Force" and recognized that U.S. position and influence in the world would be "in some direct ratio with our supremacy or lack of it, in the air." The president agreed completely that a strategic air force was the answer to the Soviet question, Lovett told Spaatz.[19]

Spaatz pressed harder for an independent air force and for the president's unification plan. He encouraged disaffection of naval aviators from their service by promising more rapid promotion in the air force. He tried to lure Radford away from the navy by offering him the top AAF command. Possibly, Spaatz instructed the Air War College to bind in the cover of an official Strategic Bombing Survey report a publicity tract that promoted the view that strategic bombing had won World War II. The AAF most likely backed publication of William Bradford Huie's sensational account, *The Case against the Admirals,* a scathing indictment of alleged Navy Department obstruction of defense unification. Over the years, Huie denied that the AAF paid for publication of his book; he later revealed that subvention came from the Arthur and Mary Lasker Fund. At the time, Stuart Symington, assistant secretary of war for air, denied that anyone in the air force had contributed either advice or money to Huie. Instead, Symington accused some naval officers of forming an anti-unification organization "through which they [flaunted] their disbelief in, and disrespect for the recommendation of their Commander-in-Chief as presented to the Congress and the people." Despite such denials, the impression remained throughout Washington in the spring of 1946 that the AAF had sponsored Huie's book.[20]

AAF activities figured prominently in Forrestal's bureaucratic tactics and relationship with Eberstadt during the remainder of 1946. Aware that Eisenhower and Undersecretary of War Kenneth Royall also thought that AAF propaganda harmed the overall goal of unification, Forrestal believed that he could settle differences personally with these like-minded men in the War Department. SCOROR worked on an outline for "possible concessions in connection with the merger issue" as the foundation for this approach. Forrestal asked Robert Patterson on March 11 if they might employ Eberstadt as the intermediary in negotiating a settlement between the War and Navy departments. He wanted the three close friends to find a common ground and to settle out of court what he referred to constantly as the unification "lawsuit," much as they had in the 1920s for corporate clients on Wall Street. Patterson now identified strongly with the single-department plan and adhered rigidly to the army's position. At first, he rebuffed Forrestal's overture: "Eberstadt's views, so far as known to me, are still the views reflected by him in his report of last fall." Patterson found this adherence to the Eberstadt Report a barrier to mediation. Nevertheless, he considered Eberstadt a close comrade since their days together in the Plattsburg training camp movement. Men like Patterson, long ago included on Eberstadt's Good Man List, worked together for a common organizational goal, regardless of seemingly bitter division over details. "Ferd might do some good work in bringing us together," Patterson admitted at last. "We both have great respect for his ability."[21]

Forrestal had opened another door for Eberstadt. On March 12, the navy secretary informed the president that Eberstadt had agreed to mediate between the two armed services. For the next six months, Eberstadt scurried between New York and Washington, meeting with the secretaries and top civilian and military representatives as they searched for a common national defense organization. During such conferences, Eberstadt listed points for discussion, arbitrated, and wrote up official summaries of the meetings. He professed complete impartiality: "From the very outset of my connection with this problem I have been interested in a balanced solution somewhere between the views of the two partisans." He devoted months to the bargaining without personal compensation. He took work home on Sundays. He flew down to Washington in stormy weather. When Truman gave the service secretaries a May 31 deadline for an agreement, Eberstadt sat up with War and Navy department representatives the entire night of May 30 to draft a statement. In another phase of the negotiations, Eberstadt brought Admiral Leahy and White House assistant Clark Clifford together on the need to adopt a "middle ground" between the two positions.[22]

Eberstadt served more than one client and negotiated simultaneously with more than one set of interests. After the Senate Military Affairs Committee voted finally to send Thomas's unification bill on to the full Senate, Eberstadt prepared a minority report for committee dissenters Styles Bridges of New Hampshire and former admiral Thomas Hart, now representing Connecticut in the Senate. Eberstadt advised Bridges, a conservative Republican, on a set of tactics that he might employ to discredit SMAC chairman Thomas. Meanwhile, Eberstadt helped the navy prepare testimony for the Senate Naval Affairs Committee hearings on unification legislation that began in early May. His influence appeared in the compilation of a witness list and in the distinctly corporatist approach of navy legislative counsel Ira Nunn, who recommended the formation of a coalition of labor leaders, journalists, scholars, scientists, clergymen, and "prominent agriculturalists" to present their views on national security to the naval affairs committee.[23]

Eberstadt took sides in the unification struggle. When Patterson refused to budge during negotiations on the issues of a single military chief of staff, secretary of defense, and department, Eberstadt undermined his credibility and bureaucratic base. He visited Plattsburg comrade Julius Adler of the *New York Times,* an Eberstadt classmate, to convince the pro-army press executive to stop Patterson from using the newspaper as a platform for his propaganda. "It seems to me that Bob has gone pretty far overboard in his recent article in the *Times,*" Eberstadt advised, "and I suspect it is under the pressure of Symington who is apparently getting rather violently active in this situation." Eberstadt persuaded former War Department officials Robert Lovett and John McCloy to bring pressure on Patterson for compromise. He told these Good Men that the secretary of war remained the only obstacle to defense reorganization and created an impasse that weakened national security at a most critical moment. The latter argument appealed to McCloy. "Not the least of my worries," he explained, "is that divided as we are, the Left-Wingers will make great inroads on our hopes for a sound military system." In the end, McCloy and Lovett intervened with Patterson, urging concessions.[24]

As Eberstadt moved among various clients and factions, facilitating, negotiating, and maneuvering, he spent less time with Forrestal. Apparently, Eberstadt failed to notice during mid-1946 what several close assistants recognized—that the navy secretary exhibited occasional erratic behavior and extreme vacillation. Once Forrestal ordered his DC-3 to turn about in mid-flight because he could not decide whether to deliver the speech scheduled at his destination. On another flight—where Eberstadt had declined an invitation to accompany his friend—Forrestal became so moody and obsessed with reading a book about the fall of the Roman Empire that he refused to speak to anyone on board the plane. He appeared tense and nervous, another assistant observed. Unification pressure buffeted him unmercifully, and at times he seemed frantic. "Get over to the President, get over to the President; tell him how wrong the Air Corps is; tell him how great the Navy aviation is; tell him this is the worst disaster in the history of the United States," one aide explained.[25]

Though remarkably patient with Forrestal's opposition to his unification plan, Truman added to the tension by placing responsibility for armed forces compromise directly upon the navy secretary and setting a series of deadlines for Forrestal to meet in endorsing the president's unification program. Each time the president requested a signed statement from his secretary, Forrestal agonized over exact wording. He directed his staff to go over every sentence time and again so that the statement would not tie Forrestal and the Navy Department too tightly to the single-department structure. Truman viewed the issue in the framework of "ward politics," alternately soothing and bullying Forrestal. In mid-June, Truman phoned Forrestal and demanded an immediate statement of navy support for an agreement with the army on unification. According to naval aide William R. Smedberg III, who sat in Forrestal's office during the telephone conversation, a visibly shaken and pale secretary hung up the receiver and said that if he did not go along, Truman would transfer naval aviation to the AAF and put the marines under the army. Forrestal said he had to tell the president he would go along with a single defense department. Smedberg considered it a case of blackmail.[26]

Forrestal counted upon Eberstadt to provide constant encouragement, intimate advice, and assistance. Sometimes Eberstadt insulated his friend from the bitter battles. He reviewed and rewrote unification statements, some rushed up to New York by naval plane so that he could study them carefully and call Forrestal about any corrections before next-day public release. In preparing one critical message that outlined Forrestal's acceptance of unification, Eberstadt revised six drafts. At the same time, he briefed Forrestal before the secretary attended various White House unification sessions and then discussed the results over the telephone as soon as Forrestal returned from his meetings with the president. When Forrestal failed to consult Eberstadt before changing a sentence in a draft letter to the president on unification, the secretary's aide, Frank Nash, apologized to Eberstadt. "I wanted you to get this word so that you might have a little understanding of why a couple of the changes suggested by you don't appear in the final result," Nash confided. "Your fuller understanding of the reasons for this will have to wait until I return to Washington."[27]

Eberstadt's mediation kept his plan at the center of the unification debate. He assumed – and only Smith and the budget bureau seemed to question – the view that all parties accepted the basic coordinating structures outlined in the Eberstadt Report as the framework for any national security law. Meanwhile, Eberstadt whittled away at the War Department structures by adjusting language that decentralized the powers and authority of a single defense secretary, military chief of staff, and department. When a British White Paper on Defense Reorganization became available during the summer of 1946, Forrestal's friend circulated copies, claiming that the British had adopted a coordinate system like that in the Eberstadt Report as the blueprint for their postwar national security organization.[28]

Eberstadt's growing influence disturbed Truman's closest advisors. Symington attacked Eberstadt. The liberal Missouri businessman served complementary political and military interests – Truman and the air force – and considered Eberstadt a dangerous enemy of both. No one seemed to control Eberstadt. When Patterson asked Symington to head the unification fight for the War Department, he declined because, unlike Eberstadt, loyalty to the administration tied his hands. "Mr. Royall or Mr. [Howard] Petersen would be much better, or a man on your staff, formally or informally, comparable to the way Mr. Forrestal has used Eberstadt," Symington responded. Nevertheless, he counteracted Eberstadt's influence. Symington spoke out for the president's unification plan and against navy lobbying and supposed intriguing. He criticized Eberstadt and his press friends as "stooges for Navy thinking." Symington represented Patterson, Truman, and the AAF in Eberstadt-moderated negotiating sessions, where he deflected the Wall Street arbiter's efforts to erode the president's single-department design and build according to the Eberstadt Plan. In one meeting, Symington confronted Eberstadt for attempting to browbeat Patterson into making a deal whereby the navy recognized air force autonomy in return for War Department abandonment of the single-defense-secretary concept.[29]

At the same time, Harold Smith at the budget office and Oscar L. Chapman, undersecretary of the interior, warned Truman about Eberstadt. Chapman insisted that the Wall Street banker intended to employ national security organization as a mechanism to corporatize American institutions. The former New Deal official contended that the National Security Resources Board and other structures gave "virtually plenary authority to fix perhaps a majority of the policies and programs of a number of departments and agencies, most notably the Interior, Agriculture, and Labor Departments." Smith agreed. He told Truman that the organizational devices in the Eberstadt Plan threatened the constitutional balance of power and democratic government. "They would bring about the most drastic changes in the top structure of the Government–affecting the President personally–that have occurred in many years," Smith said.[30]

Unlike Chapman, Smith ignored the elements in Eberstadt's organizational scheme that gave corporate advisors and outside business and industry experts direct avenues to top levels of the government. Instead, Smith framed Eberstadt's supposed threat to the New Deal institutions in the traditional framework of civil-military conflict. Perhaps the budget director used the bogey of militarism deliberately in order to dramatize Eberstadt's role. As Forrestal had noted earlier, the president seemed more upset by the introduction to government of dollar-a-year men than by the prospect of military domination over civilians. The Eberstadt Plan actually moved military matters away from the military and toward civilian experts, boards, and committees in the various coordinating agencies. In a way, the organization outlined in Eberstadt's report gave control over national defense and security to groups of militarized civilians, somewhat akin to those "Armed Progressives" who organized Plattsburg and the WIB. Both groups sought to develop peacetime structures that linked military power to national policy, business to government, and defense mobilization to the larger peacetime economy.[31]

Whatever his understanding of the Eberstadt organization, Smith's advice to the president remained the same. He urged Truman to escape from "the very awkward position of having to deal with the Eberstadt Plan" by drawing up his own presidential plan for national defense reorganization. Smith wanted Eberstadt out of the unification process. In any case, Truman preferred that his regular White House staff draw up final legislation. Apparently, Eberstadt expected to work with Clifford on the final draft, but John Kenney, assistant secretary of the navy, explained to Eberstadt,

"Notwithstanding your earlier expectations, I am confident he will not call upon you."[32]

Discouraged by the White House from further negotiation, Eberstadt's active participation in the unification battle diminished during early 1947. A job with Fred Searls, H. B. Swope, and John Hancock on Bernard Baruch's United Nations Atomic Energy Commission partly removed him from the unification scene. Private business, too, demanded his attention. "I find quite a bit of homework piled up on my desk as a result of my recent efforts to save the country," Eberstadt kidded Lovett. He limited advice to Forrestal to an occasional phone call and traveled once to testify before a congressional committee on national security reorganization, but mostly he stayed away during final bargaining over unification.[33]

Nevertheless, Eberstadt continued in numerous subtle ways to shape the final construction of America's first national security law. He guided Adm. Forrest Sherman, who, along with the AAF's General Norstad, assisted the White House staff in writing the legislation. Eberstadt clipped out sections of his report, drawing Sherman's attention to things that should be written into the bill. He warned the admiral to include both a munitions board and a resources board as separate agencies under the new structure. "There must remain constantly in being a framework or cadre to put into effect mobilization of the economic life of the nation," he reminded Sherman.[34]

In another vein, Eberstadt helped calm factions in the Navy Department, which, fearing destruction of naval aviation and the Marine Corps, grew restless as the day approached for a final unification settlement. He knew that Forrestal had obtained the best arrangements for the navy under the circumstances and hence tried to help secure naval support for the agreement. Forrestal assistant John Connor suggested that Eberstadt discuss the problem first with naval aviators Arthur W. Radford, John Towers, and Marc Mitscher, then with a conciliatory faction led by Sherman and Dewitt C. Ramsey, and finally with CNO Nimitz. "In view of the great heat that has been engendered by this discussion among the proponents of the different points of view, I think that Mr. Eberstadt will be able to get the real 'low down,'" Connor advised Forrestal.[35]

Eberstadt's mediation failed to calm the navy's fears for naval aviation and the Marine Corps. He returned to New York, leaving Forrestal alone to bring unyielding naval aviators in line with the administration's program. Forrestal assured them that their branch would be secure under the new system. Unconvinced, naval aviators pressed the secretary to carry a "Hazards of Merger" paper to the president. Forrestal also needed to appease Marine Corps insurgents who formed an anti-unification group at Quantico and lobbied vigorously against the impending legislation. Throughout the pressure-filled first half of 1947, Truman leaned heavily on Forrestal. "I think you can do more to contribute to that situation than anybody interested in it," he insisted. "I know you have difficulty with your Admirals, just as Bob Patterson has with his Generals, but Patterson, you and I," the president wrote, "are the ones who are responsible for really getting the National Defense that will work."[36]

Forrestal grew increasingly irritable. Unification was "shaking me up," he told Chan Gurney, the chairman of the newly combined Armed Services Committee. Forrestal wanted to delay hearings. When the South Dakota Republican insisted that they proceed with what he termed the "Merger Bill," Forrestal snapped back that merger was a "dirty word." During the hearings on a national security law, Forrestal grew sensitive when senators questioned his apparent sudden conversion and support of unification and implied that he had sold out. After one such exchange, Forrestal went home and telephoned Eberstadt, telling him that people acted meanly toward him. Frustrated with the attitudes of naval aviators and the Marine Corps toward the terms and language of the unification agreement and proposed legislation, he complained to former assistants Gates and Hensel that he'd had enough. After he restored morale to the service, Forrestal wrote, "I'm out of here." He told another friend, half in jest, that he wanted to quit and "turn Republican and get on some Commission or start writing my recollections which God forbid." When Philadelphia friend Richardson Dilworth suggested that he might direct the new defense establishment, Forrestal replied that under no circumstances would he do that. "If I stay here beyond the next session of Congress, I will be enjoying ulcers and bankruptcy and be wearing a bureaucrat's belly," he wrote.[37]

Forrestal's pessimism deepened as the final version of the National Security Act reached House and Senate conferees. In a luncheon on July 22 with Clark Clifford, Leverett Saltonstall, and Forrest Sherman, Forrestal revealed that there had never been a real army-navy understanding. He feared an army plot to use the new law to build a powerful single secretary who planned to merge naval aviation, the marines, and the navy into indistinct elements of one national military establishment. He doubted that the president's intention of defining roles and missions through executive agreement protected them.[38]

Despite Forrestal's constant worry throughout early 1947 (probably exacerbated by Eberstadt's absence from the unification negotiations), the final National Security Act of July 26, 1947, conformed more to the Eberstadt (and Navy) Plan than to anything else. In reality the National Security Act was the Eberstadt Plan with modifications. Every aspect of the law reflected Eberstadt's corporatist influence. Title I of the act, "Coordination for National Security," provided for the National Security Council (NSC), Central

Intelligence Agency (CIA), and National Security Resources Board (NSRB). The law gave each organization extensive (but vague) advisory powers over domestic, foreign, and military policies relating to the national security. The NSC stood as a parallel, extraordinary cabinet with authority to consider policies from any department or agency of the government concerned with national security and then to make recommendations to the president. The CIA, though specifically prohibited from police activity and internal-security functions, had advisory authority to make recommendations on the intelligence activities of all departments and agencies concerned with national security. Finally, the NSRB held advisory functions over civilian and military mobilization and specifically over stabilization of the civilian economy in time of war.[39]

Title II of the new law contained more elements of Eberstadt's organizational views. Instead of a department of defense, it called for a National Military Establishment (NME) with a secretary of defense and one secretary each for Army, Navy, and Air Force departments. Significantly, the law institutionalized the JCS as the advisory body for the NME. A Munitions Board and a Research and Development Board, both essential recommendations of the Eberstadt Report, rounded out the NME. Title III, for "Miscellaneous" provisions, although largely ignored by those who focused on the NSC and NME structures, contained a critical provision that in many ways stood at the center of Eberstadt's corporatist organization. Section 303 of Title III provided for the employment of any number of advisory committees and part-time advisory personnel. The latter served for thirty-five dollars per day or without compensation (WOC).[40]

The WOC concept embodied the idea of the dollar-a-year expert or, more specifically, the Eberstadt conception of the Good Man, who brought his corporate organizational principles to the government. The committee concept also institutionalized Forrestal's "Morrow Board" approach to national defense and security organization. Thus, for the first time there was a law providing private experts with permanent, immediate, and direct access during peacetime to the defense establishment. If the law functioned as it was constructed, through many interconnected and coordinating agencies flowing up to the president, then potentially these advisory dollar-a-year types obtained more influence on national security policy than elected or appointed government officials.

Forrestal recognized finally that the national security organization resembled Eberstadt's model. "In general, the plan that we have evolved carries with it a good many recommendations of the Eberstadt Plan," he wrote his son Michael. The corporatist forms adopted for the final law pleased Eberstadt as well. The National Security Act arose from tough, inside corporate trading, he told Robert Lovett: "There is a happy resemblance to the program that you, Jack McCloy, and I discussed in your office, and I think you are entitled to a commission on the deal."[41]

Truman responded less enthusiastically to the National Security Act. He claimed that the new law coincided with his December message to Congress. "I endorsed fully the Navy's emphasis on the need for some means of more effectively meshing military planning with our foreign policy," he insisted, "and agreed also that we needed to provide long-range plans for industrial mobilization consistent with the civilian economy." The president admitted, however, that he had accepted aspects of the Navy Plan (he never called it the Eberstadt Plan) in order to salvage legislation for a single defense secretary and separate air force department. He had obtained only partial reorganization, Truman told former navy secretary Josephus Daniels. "We can't always start out with a complete and finished organization; you must remember that since 1798 there has been a Navy Department and since the beginning of the Republic there has been a War Department," Truman explained. "It is hard to work on a bureaucracy like that."[42]

Truman surrendered after months of bitter unification debate to the Eberstadt Plan as interpreted and presented by Forrestal, the navy, and their congressional and public allies. The president abandoned, largely on Admiral Leahy's advice, the chief of staff form and accepted the JCS that he had criticized in his 1945 message to Congress. He accepted the NSC that he thought created a rival cabinet. He accepted every coordinate agency, board, and committee proposed in the Eberstadt Plan, despite obvious disdain and distrust for such corporatist entities. Truman agreed to a bill that gave easy government access to those same dollar-a-year men that he had opposed throughout the war and had criticized through his wartime Senate investigating committee. Truman biographer Robert J. Donovan observed that defense reorganization hardly resembled the president's views. Months of struggle over unification, Donovan concluded, "had wrenched the plan out of the shape originally envisioned by Truman."[43]

Still, Truman might accomplish something from his plan, if he could get Robert Patterson, his staunch single-department proponent, to head the new organization. After nearly seven years in the War Department, Patterson wanted to return to his private law practice. He turned down the president's offer of the top defense post. Then, Truman summoned Forrestal to the White House a few hours after the National Security Act had passed both houses. The president was in a hurry because he was about to leave for Missouri, where his mother lay dying. Truman came straight to the point. He wanted Forrestal for secretary of defense. There was nothing intimate nor warm about this historic meeting, only the blunt, businesslike exchange that characterized their relationship. Of course, Truman had other things on his mind, but a definite coolness and edge entered their conversation. Truman lectured his secretary-designate on how disloyal he had found Admiral Nimitz, whom he had almost fired in 1946 for indiscreet comments made at a British embassy cocktail party. He told Forrestal how displeased he had been with the stubbornness of Marine Corps commandant A. A. Vandegrift in opposing unification. On his part, Forrestal wondered aloud whether he could work with Symington, Truman's choice for air force secretary. Then, the president launched into a somewhat rambling discourse on military and naval history. Forrestal was convinced that Truman lacked an understanding of the relationship of sea power and the navy to national security.[44]

This inauspicious beginning forecast one of the most turbulent periods in American military history. Why had Truman selected Forrestal for this vital new civilian post? The two men never really understood each other, and Truman suspected this Wall Streeter's loyalty to the Democratic Party and administration. Navy secretary-designate John Sullivan believed that Truman disliked Forrestal personally. At the moment, politics and not personality motivated the president. Forrestal seemed the logical choice. He knew the inner workings of the defense bureaucracy, and no one had been closer to the entire unification controversy from start to finish. He had many key friends in Congress. Forrestal had bent just enough to Truman's views to get the presidential nod. Moreover, by selecting the head of the Navy Department to direct the entire defense establishment, Truman tied that restless and potentially rebellious service to unification.[45]

Forrestal accepted a post that he suspected might injure him financially and mentally. He responded to one well-wisher by noting: "I shall certainly need the [good wishes] and probably the combined attention of [Catholic prelate] Fulton Sheen and the entire psychiatric profession by the end of another year." He could not say no. Despite constant disclaimers, Forrestal coveted the power and position accompanying top Washington jobs. He considered it his patriotic duty and obligation to make the national security system work properly. The job had become inseparable from his life and own ambitions, and at times he thought everything would collapse without him. If American defense disintegrated because a poor secretary used his position to destroy the navy, then the United States would be unable to contain the spread of communism and expansion of Soviet power. He had created this new establishment with his friend Eberstadt and believed that they were the two people who could make those coordinate structures function. It was a "task greater than any I have ever faced," Forrestal reflected on July 29, 1947. "I only wish that Ferd were here."[46]

9. End of a Partnership, 1947–1949

The Senate quickly confirmed the popular and then widely respected Forrestal as the first U.S. secretary of defense. As a friend noted, now Forrestal had the power to lead the world toward peace or, failing in that, to build the greatest fighting machine in history. His own goals were more modest. Forrestal saw his job primarily as that of integrating the various new structures and traditional agencies into a workable national military establishment. He hoped to remove obstacles to unification, get the services to develop a common program, and build a national consensus in support of a strong, balanced peacetime military. Forrestal wanted to translate to the armed services and to the nation his own sense of duty and dedication to the national security. "There is no method, there is no way except the method of worry, of constant concern, and of unceasing energy that will give us our security," he lectured a "Team Unification" meeting attended by top civilian and military leaders.[1]

Though Forrestal probably underestimated the immensity of his task, he knew that managing an organization with a $10 billion budget and well over two million persons was the largest job that had ever confronted a federal bureaucrat. Still, it was basically a question of management, little different in Forrestal's view from his activities at Dillon, Read or his direction of the Navy Department's procurement program. It meant recruiting the best people from Eberstadt's Good Man lists to staff and operate the organization and allowing the chairman of the board of directors to coordinate programs and develop overall policy. Just as he had at Dillon, Read and as undersecretary and secretary of the navy, Forrestal asked Eberstadt to recommend Good Men and to outline the steps needed to get the organization's work under way.[2]

Partly with Eberstadt's assistance Forrestal developed a staff, although one that was severely restricted in size by the original national security act and one that budget advisor Donald Stone warned was much too small to exercise executive control. The law provided for three special assistants, and Forrestal selected Navy Department fiscal expert William J. McNeil, legal advisor Marx Leva, and John Ohly, assistant to former Secretary of War Robert Patterson. Others in the new establishment approximated the Good Man qualification. Arthur M. Hill, a Forrestal friend and Greyhound Corporation executive, agreed to head the NSRB; Eastman Kodak executive Thomas J. Hargrave directed the Munitions Board; and Vannevar Bush, the nation's leading scientific administrator, became director of the Research and Development Board. Sidney Souers began to organize the NSC as executive secretary. Lovett's appointment as undersecretary of state encouraged Forrestal, too. One of the few men close to Forrestal personally, Lovett provided that critical connection to the foreign policy establishment considered so vital to national security organization.[3]

Despite the presence of Good Men, Forrestal wanted a "driving executive secretary" like Eberstadt to make this complex and fragile corporative system function. Had Congress legislated an undersecretary of defense in the 1947 law, perhaps Forrestal would have asked his best friend to fill the post. Forrestal's own constant opposition to a strong defense secretary with a large staff prevented such action. Eberstadt probably would not have accepted a formal job, anyway. He had had enough of governmental reorganization. The White House's deliberate disregard of his advice and arbitration during final stages of the army-navy unification negotiations of 1946 had cooled him toward the task. He had expressed his disdain already for the "fool" in the White House. One Forrestal aide insisted that Truman had hurt Eberstadt's feelings.[4]

It was more likely that family matters and private business, long neglected by wartime and postwar government service, required Eberstadt's attention. Unlike Forrestal, Eberstadt pursued an active family life at his Target Rock country estate on Long Island and his retreat in the Bahamas, fishing, boating, and joining his wife in the maintenance of their prize-winning English gardens. In addition, Eberstadt refused to sacrifice personal wealth for extended public service and instead expanded his lucrative postwar aviation and Chemical Fund investments. Eberstadt avoided any official assignment in the new defense organization at the one moment that the Eberstadt-Forrestal partnership meant the most to Forrestal personally.[5]

Each day Eberstadt's waning role as advisor and luncheon partner—similar to that which he performed during the crisis-filled days after Pearl Harbor—became more significant as the secretary of defense realized the magnitude of his task. Forrestal confronted at once a problem that plagued his secretaryship, when at a "Team Unification" buffet, air force representatives went off in their own corner and barely acknowledged members of other services. Instead of showing team spirit, service secretaries—particularly Air Force secretary Stuart Symington—seemingly placed the interests of their own military branch ahead of the needs of the entire national defense establishment. This same self-interest permeated the JCS. The military, naval, and air force chiefs, each viewing issues through the perspective of his own service's roles and missions, failed to agree upon a postwar strategic plan, balanced force level, or unified defense budget.

Above all, the impatient Truman used defense reorganization to strengthen his presidency and political party. He pressed Forrestal to unify the armed forces, eliminate duplication and waste, and bring the entire establishment in line with the tight defense budget that he felt was essential to postwar retrenchment and economic stability. Truman saw his defense secretary's role not as that of a major policy advisor and national security coordinator (as Forrestal did) but in a more narrow sense, as a defense manager. Forrestal's "chief problem was that of defining specific roles and missions of each branch of the service and in determining budgetary allocations to carry out those functions," Truman explained.[6]

Confronted with such conditions, Forrestal proceeded cautiously, arbitrating disputes and persuading the competing (often feuding) groups and factions within the National Military Establishment (NME). He neither moved from his Navy Department office to the Pentagon nor held meetings with the NSC and other key agencies for nearly two months. After initial success in recruiting assistants, Forrestal found few others anxious to join the defense bureaucracy. Peacetime conditions shrank the pool of top executive talent for government as the most able of them preferred higher salaries in private business and industry. Once, eleven executives in succession turned down Forrestal's offer of a job before someone accepted. At that rate Forrestal soon exhausted a Good Man List of about one hundred names. Even those who agreed to help, such as Vannevar Bush and Thomas Hargrave, served as part-time administrators. Finally, Eberstadt told Forrestal that he had to select available people. "If you are going to wait for the best," he warned, "you may never get started."[7]

Meanwhile, news of international incidents and growing unrest inundated Forrestal's office, drawing attention and energy away from his primary organizational mission. A minor crisis over a threatened Yugoslavian invasion of Trieste actually advanced the date of Forrestal's swearing-in ceremony as secretary in September. Forrestal gleaned information about every incident and potential crisis from military and naval intelligence reports, as he had throughout the war, and now received additional information from the new Central Intelligence Agency. He also sought information and analysis from Lovett, W. Averell Harriman, and George Kennan, a State Department policy planner and Russian expert. Most of this information reinforced Forrestal's wartime suspicions that Soviet-backed or inspired communist movements would upset the peace and stability of the postwar world. Forrestal's analysts suggested that the Soviets were behind communist unrest and revolutionary movements in war-torn Central and Western Europe, China, and the Middle East. Harriman insisted that not even the American political system was immune. "Democratic leadership in the Northwest has a serious problem in handling the Communist infiltration into the party organization," he told Forrestal.[8]

Forrestal worried about the "spread of alien philosophies in our colleges" and feared communist encroachments in the Balkans, Palestine, Western Europe, and East Asia. He was not alone in this concern, as Marshall, Leahy, and most of Truman's closest advisors shared this view. Soon the president, too, became convinced that the Soviets "posed a new kind of threat to the world, because it combined a major military power and a revolutionary force bent upon conquering the world through subversion, revolutionary intrigue, and fifth columns." But Forrestal's anticommunism seemed more intense, even obsessive, and was placed in the sharp ideological context of bitter confrontation between Marxism and capitalism. During his first week as defense secretary, he began to talk about the "Russians swarming over Europe."[9]

Eberstadt tried to refocus his friend's attention, writing in early August, 1947, that Forrestal had to approach world problems "on a realistic basis." Their correspondence showed signs of the first substantial intellectual break between the two old friends. Possibly, Eberstadt felt that if anticommunist ideology drove Forrestal's conduct, he might move toward Truman's organizational views to combat the perceived communist menace rather than build the corporatist foundations that Eberstadt felt lay at the base of their national security partnership. Eberstadt grew impatient with his friend's ideological views. Communism did not mean the end of civilization, he told Forrestal, saying that "in human events there is no last chapter." Eberstadt thought that the United States should build its own sound political-economic system–especially those structures in the national security organization that linked the civilian and military, government and business sectors–rather than crusade against the spread of communism through military intervention or expensive "doles" to every country threatened by revolutionary unrest. "Unless we can get agreement on a sound and promising plan of recovery," he wrote Forrestal, "it might even be better for us to let non-producing countries go Communist if they want to and thus throw their burden on Russia, than for us to attempt permanently to hire them to wear a dubious democratic label by pouring our resources down the drain."[10]

Despite Eberstadt's admonitions, Forrestal pursued his interest in Marxist ideology and Soviet activity. He expanded earlier studies made by E. F. Willett, George Kennan, Robert Strauss-Hupe, William Yandell Elliott, and others. He encouraged Kennan to prepare an essay, "The Sources of Soviet

Conduct," that was published under the signature "Mr. X" in the prestigious journal *Foreign Affairs,* an essay generally credited with launching the American Cold War containment policy. Forrestal solicited other analyses from academic, business, and government sources. He paid particular attention to a somewhat oversimplified response from the American naval attaché to Moscow, L. C. Stevens, a naval aviator. Stevens explained that an imported Germanic ideology developed by Lenin drove current Soviet policy and actions. He told Forrestal that the Soviets had assembled this Marxist-Leninist ideology in a little primer so that they could spread their subversive philosophy to what Stevens termed "every dim mind" in the world. "The men in the Kremlin have made false philosophers out of a large part of the world, but the falsity of their philosophy can and should be clearly shown," he wrote. This could be accomplished, the attaché continued, by "one of our research foundations" compiling a simple, convincing study of the free-market system. "I believe that such a counter-effort on our part is necessary for our own people," Stevens explained, "particularly the muddled liberals and the younger generation."[11]

Forrestal shared Stevens's letter with Princeton University president Harold Dodds and with Eberstadt. He told Dodds that many American universities–most notably Harvard–had swallowed this dangerous German-Marxist philosophy. Forrestal found that American labor, and "most of our people," had fallen under its influence as well. Only conservative economist Friedrich A. Hayek, whose book *The Road to Serfdom* had just been excerpted by *Reader's Digest,* impressed Forrestal with his understanding of how increased state control and central planning led to muddled thinking and down the path to socialism and communism. Forrestal wondered if someone on the Princeton faculty might develop Hayek's ideas in a capitalist primer to counteract the Marxist primer that Stevens had talked about. Forrestal wanted a book that he could hand to American labor and members of "other classes" to educate them to the dangers of communist ideology.[12]

Eberstadt studied copies of this correspondence and seemed appalled by the simplistic approach that his friend was taking toward the complex issues of the postwar world. No one could stimulate the spontaneous creation of a capitalist tract, Eberstadt argued; and the United States had yet to produce a Rousseau, Toynbee, or similar great mind capable of undertaking the broad philosophical and historical synthesis proposed by Forrestal. Eberstadt focused instead on pragmatic organizational solutions with which to meet postwar problems. The United States needed realistic plans, programs, and structures to handle "the stress and strains of the present readjustment," Eberstadt instructed Forrestal. "The question is to discover how we can purchase security at a reasonable cost."[13]

Eberstadt's dialogue with his best friend over the capitalist primer suggested a further subtle but perceptible breakdown in their common understanding of the national security partnership, or at least what Eberstadt felt was the basis of that partnership. In response to Forrestal's search for a capitalist manifesto, Eberstadt reminded his best friend about what constitutes the real foundations of national security, underscoring the close relationship that he saw between unification and corporatism. Eberstadt wanted Forrestal to concentrate on the development of the coordinating machinery that they had articulated in the Unification Report of 1945 and helped build into the law of 1947. Those structures offered one way to forge a system of agencies that coordinated and harmonized the public and private sectors.

Eberstadt maintained that through agencies such as the NSC, NSRB, and the Munitions, Research and Development, and Education and Training boards, organizational experts could rationalize the political economy by "constant careful peacetime planning and preparation." They could hold industrial mobilization maneuvers that paralleled military training exercises. National security structures could stimulate military contracting to stabilize the economy, stockpile raw materials, and allocate national resources among various claimant agencies that made up the entire economy. Under unification, the national security organization could "exercise such proper influence as it can toward inducing manufacturers to install the very latest and most up-to-date equipment," Eberstadt explained to Forrestal. "It would be good business to do so, as well as good national security."[14]

Eberstadt agreed to come down to Washington once more so that he could help get the national security machine functioning. Between late August and early November, 1947, he visited Forrestal regularly, including attendance with a select few invited guests in September at Forrestal's swearing-in ceremony. "I am spending about a day a week with Jim, kibitzing on his new organization," he confided to Lovett. As usual, Eberstadt prepared a Good Man List that included, ironically, the name of former Assistant Secretary of War Louis Johnson, who would succeed Forrestal as secretary of defense in early 1949. At the same time, Eberstadt reviewed organizational charts for the Office of the Secretary of Defense (OSD) drawn by management consultants Cresap, McCormick and Paget. He checked Forrestal's speeches and statements on the progress of the National Military Establishment. Eberstadt also evaluated proposed new weapons systems that would stimulate American industry, particularly a program to develop the atomic submarine. "As I have already written you," Eberstadt advised, "I feel sure that the undersea arm of the Navy will certainly be one of its most important and possibly its most important weapon."[15]

Eberstadt served temporarily as Forrestal's troubleshooter, checking rumors and reports of alleged disloyalty to the secretary from the air force and the army, which both appeared indifferent to the secretary's pleas for a unified budget. Eberstadt advised his friend to take charge, "as the new National Security Act puts major budgeting responsibilities on your shoulders." Eberstadt joined Baruch in evaluating the organization of the National Security Resources Board. They considered this agency as the key mechanism to connect unification to a larger corporate political-economic organization by coordinating military, industry, labor, and business in a national security program. Drawing on his War Industries Board and Army Industrial College experiences, Baruch pressed for the NSRB to obtain authority over the collection and control of all financial, economic, and industrial information necessary for the development of an industrial mobilization plan. Baruch urged Eberstadt to teach basic corporatist principles to NSRB chairman Arthur M. Hill, "or he may find this 'cold war' developing into a pretty warm one." Baruch reminded him of the previous postwar era when the nation abandoned its corporatist experiment: "I find the same basic differences commencing to develop in the set up, as happened after the War Industries Board, only they are developing faster."[16]

Baruch and Eberstadt recommended that Hill obtain presidential authority for the collection of economic information from other executive departments and take charge of mobilization planning for the entire government. Eberstadt explained the philosophy behind this recommendation: The armed services and the Munitions Board demanded their own share of the national resources without thought of the needs of other parts, while the NSRB provided what he insisted was an impartial control mechanism, somewhat akin to the way the WIB functioned in World War I. Both judged how the allocation of resources to each service and other claimant agencies affected the entire economy and the national security. However, this approach to the NSRB met immediate opposition from the service secretaries and from the White House, and the ensuing debate over the board's functions and authority focused on the entire process of postwar adjustment in the national security machinery.

Behind opposition to the Eberstadt-Baruch idea of a greatly strengthened NSRB stood the old New Dealers' suspicions that such a corporate board would become the platform for a corporatist managerial elite. Supposedly this elite planned to redesign the state, subvert presidential and congressional authority, disrupt the constitutional balance of power, and undermine democratic government. Truman remained particularly uneasy with the type of organization embodied in the NSC and NSRB. He largely ignored the NSC and never sat in on its deliberations as provided for in the 1947 law. He thought the NSC and NSRB fit a European model of cabinet government rather than the American elective system. These agencies sought collective responsibility through a select elite rather than responsibility by the elected representative of the people. "Under our system the responsibility rests on one man—the President," Truman argued. "To change it, we would have to change the Constitution, and I think we have been doing very well under our Constitution," he wrote.[17]

When Forrestal presented to the Cabinet the idea of reorganizing and strengthening the NSRB, he discovered strong opposition. Secretary of State Marshall, intensely loyal to Truman and the president's conception of government, voiced immediate reservations. Marshall revealed a connection between the methods and goals of the earlier New Era corporate organizers and the present crowd. He compared the Forrestal-Eberstadt objective of strengthening the NSRB to Charles Dawes's attempt in 1921 to make the budget office he directed into the office of a "second president." During that earlier era, Dawes had demanded control over all information coming into the executive, just as Forrestal and Eberstadt now proposed in their NSRB scheme. Army secretary Royall supported Marshall and insisted that the NSRB remain nothing more than an advisory agency. He wanted the Munitions Board to continue, as it had since the 1920s, as the chief point of contact between military and private industry and the agency for mobilization planning.[18]

Truman decided in November, 1947, to grant the NSRB limited authority to request information. He kept the board as an advisory panel. Eberstadt accepted the decision. He had left an outline for Hill to get the NSRB "rolling," so that in a national emergency it could develop the framework for full mobilization. "The basic consideration is to put the burden as near as possible to where it will fall in case of actual mobilization," he told Forrestal, "so that the work will be not merely a study, but will in fact be preparation for an emergency." Finished with the first stage of his "kibitzing," Eberstadt joined Patterson on a tour of Europe to evaluate conditions for rebuilding the economy of Western Europe under the Marshall Plan. Patterson traveled as the codirector of a Citizens Committee for the Marshall Plan and warmly endorsed the idea of a massive infusion of American foreign aid to help rebuild Europe. Not so Eberstadt. He worried that the plan meant bigger government and a more expensive and intrusive federal bureaucracy. He preferred the quasi-official Dawes Plan of the 1920s with its board of private experts and transfer mechanism that adjusted conflict and coordinated private business with government policy.[19]

Back in Washington, Forrestal continued his battle to organize the National Military Establishment and to gain service cooperation in the development of a unified budget, strategic plan, and balanced defense force. It was a bitter fight that drained the secretary. David Lilienthal, chairman of the Atomic Energy Commission, found Forrestal in late 1947 "looking pretty worn," while another visitor to the secretary's office heard Forrestal rave about the employment of atomic energy as a weapon rather than as a resource to be used for peaceful purposes and considered the secretary to be a dangerous man. Certainly he was a troubled man. Forrestal worried constantly about military preparedness, budget cuts, service efficiency and morale, and how to keep the government and world out of the clutches of what he called "intellectual muddlers, crackpots," and "our friends to the east." Forrestal attended to wide-ranging military, diplomatic, and espionage activities, including possible U.S. involvement in covert operations in Greece, Italy, France, and China. As he explained, he constantly had to "tie up loose ends." Whether Forrestal operated through a profound understanding of power as the basis for national policy or through simple ambition to be involved in every critical issue, he spread his interests everywhere, much as he had as navy secretary. Simultaneously, he tried to develop the defense structure, balancing innovative new security agencies with the traditional military and political systems.[20]

Forrestal's approach to the "Palestine question" was typical of his approach to the complex international situation confronted by the Truman administration between 1947 and 1949. His understanding of the region's significance derived from years of pursuing Dillon, Read oil syndicate investments in the Middle East. Forrestal considered access to oil vital to American economic and strategic interests and consequently recommended an even-handed approach to the growing struggle between Arabs and Jews over the establishment of Israel. Forrestal believed that controversy over formation of this independent Jewish state inflamed the situation and hurt the American strategic position in the Mediterranean region. He wanted to take Palestine "out of politics." Instead, Forrestal drew the ire of pro-Israel elements in his own Democratic Party and throughout the country, making him more vulnerable to the concerted and venomous public criticism that had started to plague him as defense secretary.

These attacks on Forrestal arose from a number of sources. Henry Wallace accused Forrestal of continued association with German industry that had supported Hitler and consequently opposed Forrestal's selection as defense secretary. Other liberal Democrats saw Forrestal as a dangerous militarist who wanted to give the armed forces control over atomic weapons and blocked any reconciliation with the Soviet Union. For their own reasons, Symington and friends of a strong air force also spoke against Forrestal. Drew Pearson and other journalists reported scandalous items about Forrestal's personal and business connections and conduct. The Palestine issue merely intensified this increasingly bitter anti-Forrestal campaign. Eberstadt observed that the charges of anti-Semitism and innuendo about Forrestal's sympathy for fascism deeply bothered his friend. Eberstadt thought the charges most unfair. "I know of no more truly democratic or unprejudiced man than he is," Eberstadt wrote.[21]

Strangled by Washington politics and press, Forrestal now learned that U.S. military forces had declined, through demobilization and postwar budgetary cuts, far more rapidly than he had expected. When Army Chief of Staff Eisenhower retired in February, 1948, he handed the secretary a farewell memorandum that revealed glaring weaknesses in the army. Eisenhower's message enumerated dangerous shortages in manpower, weapons, and equipment. "This is a good paper and each word has been soundly based," Forrestal told Eberstadt. Forrestal's most trusted military aide, A. M. Gruenther, added his warning. Gruenther observed that a weakened defense establishment severely limited American ability to project military power overseas. The final shock came on February 12 at an NSC meeting in which Marshall explained that as the government increased its commitments to contain communism throughout the world, it decreased the forces necessary to implement those commitments. The United States was "playing with fire while we have nothing with which to put it out," he warned.[22]

Forrestal blamed himself for these developments. He worried that his inability to bring the armed services together on a common strategic plan, budget, or program contributed to this dangerous situation. Immediately following the depressing NSC meeting of February 12, Forrestal employed Title III of the National Security Act once more and asked for Eberstadt's help. "I do not like to make too many demands upon your time, as you know, but I think the growth of the child of your brain is by now so far advanced that it would be interesting to you, and certainly profitable to me, to have you take a look at its development whenever your time permits," Forrestal wrote. No one in the government could play the role that Eberstadt served. Though he remained close to Robert Lovett, Forrestal knew that the undersecretary of state owed his obligation and duty to Marshall. Forrestal developed no intimate ties with Kenneth Royall, Stuart Symington, Dean Acheson, George Kennan, or anyone else in Truman's policy establishment comparable to his wartime bonds with Robert Patterson, John McCloy, W. Averell Harriman, and Harry Hopkins. He was not close to Truman or to any of the White House staff.[23]

Eberstadt arrived at the Pentagon in late February, 1948, where he reviewed Eisenhower's retirement statement and a draft of Forrestal's first annual report for what was informally known as the Department of Defense. The report suggested that Forrestal recognized structural (but not overall conceptual) weaknesses in the NME. He wanted the addition of an undersecretary, a larger joint staff, a chairman for the JCS, a personnel board, exclusion of the service chiefs from NSC meetings, and a greatly clarified definition of the secretary's powers. Eberstadt recommended the attachment of a controller and closer cooperation between agencies responsible for the preparation of mobilization plans and the research and development of weapons. Eberstadt believed that Forrestal had made progress in the organization of the department. Despite problems, Forrestal's office had developed a uniform pay scale, unified budget, and coordination and inventorying of procurement functions. The Office of the Secretary of Defense had studied reserve forces, a unified medical service, and reviewed civil defense arrangements. Above all, Eberstadt found a continued sense of mission and intense loyalty among Forrestal's assistants.[24]

Neither Eberstadt nor Forrestal questioned the basic format of their national security organization. They expected to remedy flaws in operations by modifying institutional structures. In their view, problems arose from the inability of the JCS or individual armed services to agree upon a strategic plan that would trigger the larger process of economic control through appropriations, contracting, and government-industry cooperation. Possibly, Eberstadt never recognized that the JCS and three military services could not function like a corporate board of directors driven to agree on policy by the demand to achieve maximum production and profit. He thought that the JCS and military could be disciplined and forced to conform to the corporatist scheme. He told Forrestal that the JCS betrayed public faith in the military and gave potential enemies the satisfaction of knowing that the United States could not back up its foreign policies with military force. "It seems to me someone ought to talk to these gentlemen like a Dutch Uncle," he advised Forrestal. Following this suggestion, Forrestal arranged a meeting for early March with the Joint Chiefs of Staff at the Key West naval base, in order to hold what the secretary called a "frank discussion."[25]

As he finished his review, Eberstadt sensed—possibly for the first time—that his best friend showed unusual strain. Forrestal's careful reflection on issues (some called it indecisiveness) had become more cautious and sensitive than before. Eberstadt never blamed problems with the national security organization for Forrestal's breakdown in health. He found other reasons. "My friend has apparently gotten himself in very wrong with the Zionists," Eberstadt observed as he left Washington for a vacation in the Bahamas. That same day, news arrived that the communists had toppled the government of Czechoslovakia, and Forrestal was deeply disturbed about Czech patriot Jan Masaryk plunging to his death from his office window. The information forecast further trouble and promised to cut short Eberstadt's vacation, bringing him back to Forrestal's side. The communist takeover initiated a series of events known collectively as the March Crisis of 1948. A report from Gen. Lucius Clay in Berlin that war with the Soviet Union might come with "dramatic suddenness" increased the tension. Truman responded with a belligerent St. Patrick's Day address to Congress, calling for reinstatement of the military draft, passage of a universal military training law, and funding of a massive foreign aid program under the Economic Cooperation Act. Truman's speech catapulted Washington into a self-conscious atmosphere of crisis unknown since the end of the war.[26]

Truman undoubtedly saw a threat to American security from developments in Central Europe, but he also employed the crisis as a calculated political move to gather public and congressional support for programs and appropriations. His hard line toward the Soviets owed something to Forrestal's constant advice to contain communism, but the overall tone of the message caught the secretary of defense unprepared to present the administration's military requirements to Congress. Forrestal watched as Symington and air force spokesmen pressed for the funding of a seventy-group air force as the mainstay of America's defense, although the defense secretary held that such a program gave "an unbalanced military organization and an illusory sense of security." Forrestal recognized that the American military establishment was altogether unprepared to fight another great war in Europe. It lacked adequate manpower, weapons, supplies, and, most important, a comprehensive strategic plan. In the end, Forrestal thought that the United States might have to use atomic bombs. Consequently, he counseled caution during a crisis that included the dispatch of B-29 heavy bombers to Europe and a great airlift of supplies over a communist blockade of Berlin. Forrestal's apparent timidity and indecisiveness angered Truman, who began to feel that his secretary "passed the buck, hedged, and alibied" when it came to military advice.[27]

Forrestal's deliberate approach arose partly from his concern that the NME and security organization should function as they were designed under the National Security Act. CIA, JCS, and NSC needed to provide information, analysis, and advice on the situation in Europe before the secretary and president acted. The NSRB had to prepare a mobilization plan. Forrestal required assistance for this task, and once again crisis drove him to seek Eberstadt's counsel. Two days after Truman's St. Patrick's Day speech, Forrestal dispatched an urgent cable to Eberstadt in Nassau, requesting that he return immediately to Washington to help the NSRB construct its mobilization plan. Though reluctant to become entangled in the politics of mobilization and probably suspicious of Truman's actual commitment to planning, Eberstadt agreed to assist.

Eberstadt recognized another chance – perhaps his last one – to rationalize what he considered the basic coordinating structure to link national economic and security policies through the NSRB mobilization plan. Eberstadt renewed contacts with leading military and civilian mobilizers such as Brehon Somervell, of Koppers Corporation, who were now prominent in private business and industry. It had been Somervell, then chief of the Army Service Force, who at the end of the war had exhorted Eberstadt to continue the work that they had begun during the national emergency. "I hope that you can carry the banner in seeing that some of these things which we all want so much can be actually made to materialize," Somervell had instructed Eberstadt in 1944.[28]

As a consultant to the NSRB (hired under Title III of the National Security Act) in April, 1948, Eberstadt determined to carry that banner into the postwar era. Once more he assembled the Good Men who had helped him with the Munitions Board, CMP, and unification studies between 1940 and 1945. Munitions Board ally Stacy May, now with the Rockefeller Foundation, studied an economic stabilization program that could be implemented under a plans and programs division of the NSRB. Eberstadt consulted once more with former wartime associates Bert Fox, John Fennelly, and Milton Katz. He talked to Baruch, John Hancock, and Fred Searls. He worked with Simon Kuznets, a Wharton School of Economics professor, and economist Edward G. Nourse, a Brookings Institution vice-president and chairman of the Council of Economic Advisors. Recognizing that the Economic Cooperation Administration under the Marshall Plan had become a major new claimant agency contesting for resources with the military services and domestic economy, Eberstadt held a planning session with ECA administrator Paul G. Hoffman.[29]

Secretary of the Interior Julius Krug, who Eberstadt called one of his boys from the War Production Board, seemingly lent administration support for the development of an NSRB mobilization blueprint. Krug advised that the NSRB organize a statistical service and collect industrial, labor, and energy information. He recommended development of a priority rating system for critical materials and stockpiling preparatory to the reinstitution of the Controlled Materials Plan. "Failure to do this in time," Krug warned, "will lead to the same helpless confusion, unbalanced inventories, and erratic production which characterized the first two years of the emergency we have just lived through." Urged on by Krug, Eberstadt expanded his mediation, linking the NSRB to the NSC through old study group member Sidney Souers, now NSC executive secretary. He then tied these agencies to other segments of the national security system through personal meetings with John McCone, Elihu Root, Jr., Allen Dulles, and other national security experts. These conversations had one immediate result. Forrestal hired Dulles—Eberstadt's classmate, yachting companion, and William J. ("Bill") Donovan's wartime OSS agent—to review CIA personnel, operations, and an intelligence product that disappointed Forrestal.[30]

Eberstadt's presence and tinkering with national security machinery concerned the Truman administration. Budget director James E. Webb lectured Eberstadt about his responsibility to make the NSRB report and mobilization planning conform completely to the president's thinking on the subject. Truman viewed the board as an advisory body without independent authority. The president opposed any attempt, such as that suggested by Eberstadt, to create an "alternate" source of power, particularly one where a collective group of nonelected outside experts from private business associated with government bureaucrats to conduct quasi-official government security policy. In some ways, the NSRB debate mirrored WPB controversy during the second world war, when two different conceptions of the role of government in the political economy had clashed. Like the New Dealers, Truman's Fair Dealers feared that Eberstadt sought to dismantle government regulatory and social welfare machinery and replace it with corporatist agencies such as the WPB or NSRB that bypassed regular executive and legislative channels. In this scheme, regular government became little more than an umpire overseeing policy already determined by a small, privileged elite.[31]

Eberstadt understood that the politics surrounding his NSRB review symbolized a much larger political-economic struggle and conflicting view of the state. He compared the situation to the crisis in 1940, when FDR had convened the War Resources Board to appease proponents of preparedness, directed the compilation of an industrial mobilization report, and then ignored the board's work. Now, Eberstadt suggested that another big-government, New Deal politician was using the NSRB report to increase executive power and obtain billions for the Marshall Plan, atomic energy program, and "more billions for military and perhaps more billions to come for lend-lease." By stimulating and then rejecting his plan, in Eberstadt's opinion, Truman had unleashed a military-industrial combination. "The effect of denying the Board any responsibility with respect to coordination opens the door to the imposition, without control or coordination, of the huge burden of security contracts on an economy already heavily laden and, in many sectors, severely overextended," Eberstadt explained to Forrestal. "Except for such restraints as the contracting agencies may impose upon themselves, they are subject to no formal limitations," he wrote.[32]

NSRB legal counsel Kenneth Johnson noted Eberstadt's growing disillusionment, particularly over the president's decision to transfer mobilization planning from those on the Good Man List to military-industrial interests and big-government bureaucrats. Eberstadt had suffered a major setback. In the words of Robert Cuff, a national security scholar, "the kind of coherent, centralized vision that Eberstadt cherished cracked apart on the competitive pluralism of the executive branch, and the determination of President Truman to protect his personal autonomy in the decisionmaking process." Johnson hoped that Eberstadt would continue the fight for his vision of national security. "Certainly, as the 'putative father' of National Security you can never forsake your offspring," he assured Eberstadt. Nor would Forrestal and former president Hoover allow Eberstadt to forsake his "offspring" as they arranged for him to head a Task Force on National Security Organization as part of Hoover's Commission on Organization of the Executive Branch of the Government.[33]

The Hoover Commission seemed the perfect format for Eberstadt to continue his organizational intermediation. Conservatives in Congress had created this body of experts to prepare an anti–New Deal, anti–big government blueprint for the new Republican administration that most expected to replace the struggling Truman government in the upcoming election of 1948. This blueprint would undoubtedly reflect the vision of commission leader Hoover and his staff, which included longtime Eberstadt associates Owen Young, Henry Stimson, Donald Davis, H. Struve Hensel, John Hancock, and New York public planning expert Robert Moses. Most important, "The Chief," as Eberstadt addressed Hoover, dominated the group. Hoover's theories of an associative state and corporatism came closest to Eberstadt's own views. Eberstadt praised the former president for his lifetime struggle for "freedom and opportunity against government restrictions and the ambitions of power-hungry individuals."[34]

Despite the favorable conditions, Eberstadt hesitated to join the Hoover Commission. Perhaps the structure of the study group bothered him. The legislative act called for the president, the speaker of the House, and president pro tempore of the Senate to appoint two members each from private and public life. These twelve members would form a bipartisan panel. But Eberstadt had had enough dealings with Truman over unification and most recently in the NSRB study to recognize that the president could turn such a commission to his political advantage and have it endorse big government and New Deal institutions. Even Truman's appointment of Forrestal (at Forrestal's request) to the commission did not calm Eberstadt's concern. Though he never told Forrestal directly, Eberstadt believed that his best friend, under pressure from the president, was moving toward the Truman-Marshall conception of unification with a powerful secretary and chief of staff. Moreover, Eberstadt worried that Hoover had converted to the president's views on reorganization. "I told him . . . that if he had his mind made up that the three military departments should be thrown into one," Eberstadt explained, "I was the wrong man to get and would not commit myself in advance to any such conclusion."[35]

Eberstadt agreed finally to head the national security task force, only if Hoover broadened the scope of his investigation to include not only the National Military Establishment but the entire national security organization. Hoover agreed to Eberstadt's terms. This drew an immediate reaction from air force secretary Symington and army secretary Kenneth Royall. Neither trusted Eberstadt. Symington's opposition appeared personal, but Royall's stemmed from concern that Eberstadt planned to block further defense reorganization and executive centralization. As Royall told Hoover, "I believe that experience has completely demonstrated that Mr. Eberstadt's original views on unification are largely fallacious, particularly in failing to realize the necessity for the existence and exercise of a strong central authority in military and logistic matters." Hoover ignored Royall and warmly defended Eberstadt. During the course of the study and the difficult period of criticism following Eberstadt's work, Hoover showed a steadfast faith in and loyalty toward Eberstadt that blossomed into a warm friendship between the two men somewhat akin to the Baruch-Eberstadt relationship. The former president recognized a kindred corporatist.[36]

Supported by Hoover and guided by Forrestal until he began to suffer severe health problems toward the end of the commission's deliberations, Eberstadt recruited his Task Force on National Security. Composition of the team revealed the larger policital-economic motivation and organizational scheme embodied in the Hoover Commission. Eberstadt adopted the model of the Morrow Commission of 1925 by assembling an elite group of experts, including former government officials, business leaders, educators, and journalists. There were three university presidents, two leading journalists, a top General Motors executive, an influential aviation executive, and an atomic energy expert. Eberstadt selected two marketing experts with extensive war mobilization experience. Wartime allies Patterson and McCloy joined the group, as did Chester I. Barnard, president of the Rockefeller Foundation. Eberstadt, McCloy, and *New York Times* correspondent Hanson Baldwin made up the Task Force Steering Committee, while E. F. Willett, Eugene Catron, and Robert Howe Connery–reprising their 1945 Eberstadt study group roles–headed a twenty-five-man support staff. Morrow Board veteran Jimmy Doolittle and eight retired officers formed a Military Advisory Board, while Hancock, Eisenhower, Admirals Towers and Nimitz, and six other wartime leaders served as task force consultants.[37]

The Eberstadt Task Force on National Security held 59 sessions between June 8 and October 20, 1948, interviewing 245 witnesses in person. Eberstadt spoke with scores more on the telephone. "I recall that when we were doing the Hoover study," Connery wrote in 1970, "we had to put another operator on duty just to handle his long distance calls." The sheer mass of information collected and tremendous number of people interviewed obscured the fact that Eberstadt employed his task force to endorse and institutionalize the corporatist forms and structures presented in his unification report of 1945 and adopted as the framework for the National Security Act of 1947. By employing the supposed threat of JCS and individual department militarism, Eberstadt's team recommended the strengthening of the CIA, the civilian advisory board network, and the planning apparatus for industrial mobilization. They brought the JCS and military department firmly under civilian control, Eberstadt concluded, because the "proper relationship of this new force to our political and economic fabric can make the difference between democracy and totalitarianism."[38]

Eberstadt's conclusions brought instant criticism from those who wanted further centralization and the addition of a single military chief of staff. Patterson refused to endorse the task force findings and called the report a whitewash. "We might as well face up to the fact that inter-service frictions have not been reduced, and that costly duplications have not been diminished," he complained to Eberstadt. Symington attacked both the committee and the report, calling them pawns of the Navy Department. Task force member Hanson Baldwin received a severe rebuke from the air force secretary. "Symington indicated that if the Committee did 'publish the Navy report,' or make recommendations on the subject, he wouldn't want to, but might have to, 'knock down' our reports," Baldwin reported to Eberstadt.[39]

The criticism bothered Eberstadt. In his own mind, he had not slanted the task force in order to defend the navy in its ongoing struggle for power, weapons, or budget with the air force. "Our only client was Uncle Sam," he told an ABC radio audience. His differences with former army unifiers such as McCloy and Patterson were not great, he assured journalist Arthur Krock. Eberstadt wrote Eisenhower, explaining that he had given each service an equal opportunity to testify and present its view before the task force board. In reply, Eisenhower praised Eberstadt. He insisted that the nation owed him a debt of gratitude for his "disinterested desire to serve the welfare of the country." Though privately Eisenhower felt that Eberstadt paid only "lip-service at the altar of unification," he assured the task force chairman that he had shown nothing but a common-sense approach to the national security issue.[40]

Eisenhower's personal expression of confidence pleased Eberstadt, who began to consider the former general to be the type of statesman that he had sought to lead postwar America. Eberstadt promoted "Ike" to J. F. Dulles, the Luces, and Hoover as a possible Republican Party presidential candidate. He contributed money to organize an Eisenhower-for-president movement. As he told Forrestal, "Ike" was dedicated, unselfish, and "the very best type of American." Possibly Eberstadt recognized Eisenhower's sympathies for a political-economic system that historian Robert Griffith identified later as Eisenhower's "corporate commonwealth." Eisenhower shared Eberstadt's alarm with big government that exerted "ever greater power over our daily lives" and with federal regulatory agencies that "nag, irritate, and hound every businessman in the U.S." Most important, Eisenhower searched for the foundations of a stable American foreign policy by meeting and studying with the members of the Council of Foreign Relations who had been included over the years on Eberstadt's Good Man List, men such as Allen Dulles, John Cahill, Stacy May, Russell Leffingwell, and George Small Franklin, Jr., the son of Eberstadt's former law partner at Cotton and Franklin.[41]

Eberstadt's admiration for Eisenhower became more significant during the latter part of 1948, as the former army chief of staff arrived at the Pentagon to assist an increasingly beleaguered Forrestal. "I'm glad you will help our friend who is in trouble," Eberstadt wrote Eisenhower on December 7, adding that "his greatest need is for competent people, loyal to him and to the overall national objective." Eberstadt recognized at last that his best friend's recent bout of pessimism differed from those familiar periods of gloom followed by elation that Eberstadt had observed in Forrestal over the years. He had been so close, understood Forrestal's moods so well, that he ignored signs of eventual breakdown. Forrestal never revealed inner torment to Eberstadt until the very end in 1949. He remained calm and forceful in Eberstadt's presence, maintaining what might be viewed as the Good Man's tough façade of an unemotional corporate manager. In reality, Forrestal was losing his grip. Eisenhower concluded, after meeting with the defense secretary, that "he blames himself far too much for the unconscionable situation existing now." Budget director James Webb was less charitable. He criticized Forrestal for allowing the military to spin hopelessly "out of control." According to Webb, Forrestal refused to take responsibility any longer, forcing the president to act as his own secretary of defense. Reportedly, Forrestal sat with his head on his desk, sobbing about his failure to present the president with a satisfactory defense budget. Eberstadt observed in a note to Hoover that they were "dealing with a different person than the man that I have known for 30 years."[42]

Forrestal fought on in the only way he knew how. He continued to negotiate and persuade feuding factions in the defense establishment while searching for Good Men to manage affairs. He called out of retirement Eisenhower, Spaatz, and his World War I boss, Adm. John Towers, to help him get the services together on a unified strategy and their respective roles and missions. Forrestal took the Joint Chiefs to Newport, Rhode Island, where they talked compromise over whether the navy or the air force ought to control atomic weapons. But after Truman's surprising reelection in November, Forrestal's ability to manage declined markedly. He visited the president, hoping to define both his position in the new administration and Truman's proposed military policy. Truman ignored him, leading Eberstadt to advise that Forrestal resign at once. "That was tip-off enough for me" Eberstadt wrote, "and I suggested that he draw the obvious conclusion, which he was reluctant to do."[43]

Instead, Forrestal hoped that the retired army general could save the defense establishment. Eisenhower observed that the secretary assumed that the general had "some miraculous power to make some of these warring elements lie down in peace together." Eisenhower added that Forrestal "exaggerates greatly the possibility that I will materially help in his task of 'unifying' the services." With Eberstadt out of the country between December, 1948, and early February, 1949, Forrestal relied on Eisenhower, much as he had on Eberstadt, to provide counsel and comfort. Eisenhower tried. He recognized immediately that Forrestal suffered. "Jim is looking badly," the general noted on January 8. Eisenhower encouraged Forrestal to focus on the problem of strengthening national defense organization by implementing recommendations presented by the Eberstadt task force. He kept the defense secretary's mind on his managerial work.[44]

The situation for Forrestal continued to disintegrate. Truman offered no help and in fact seemed "damn sore" with Forrestal for placing him in the middle of the interservice squabbles that gave Republican opponents the chance to attack the administration. Moreover, Forrestal had not worked very hard for Truman's reelection. Apparently, close Truman friend and advisor Harry Vaughan and financial backer Louis Johnson intrigued to remove Forrestal. Personal attacks continued to appear in the gossip columns of Washington's newspapers. In the meantime, neither Forrestal nor Eisenhower made much progress toward bringing the services together. Eisenhower found that navy and air force differences over budgets, strategic plans, and whether the long-range B-36 bomber or the super aircraft carrier should become the backbone of America's defense had become bitter, "personal" struggles for influence and power. Sadly, Forrestal told Eisenhower that both services had betrayed him. "At one time he accepted unequivocally and supported vigorously the Navy 'party line' given him by the Admirals," Eisenhower suggested. Now, he trusted only Admirals Forrest Sherman and William H. P. Blandy. "It must have cost him a lot to come to such a conclusion," Eisenhower observed.[45]

By early February, Eisenhower expected something to snap in the defense establishment. It was Forrestal. Returning to Washington on the evening of February 15, Eberstadt visited his comrade at home. He found Forrestal deeply depressed. Eberstadt insisted that Forrestal quit at once and take a long vacation. Then, after rest, he could return to accept an ambassadorship to France, England, or the Soviet Union. As to vicious press attacks over the past months by Drew Pearson, Walter Winchell, and other journalists, Eberstadt advised against any public response. Forrestal could take legal action later. Forrestal seemed not to listen. Two days later, Eberstadt tried again. He thought Forrestal might visit Baruch's South Carolina country estate for relaxation and to learn from the "Boss" who was behind these terrible attacks on the secretary. Forrestal ignored this advice. Next, Eberstadt persuaded classmate and Texas oil millionaire Thomas Armstrong to invite Forrestal down to his ranch. "Jim was in a rut" and needed to get away, Eberstadt wrote. Still, Forrestal refused to leave his desk.[46]

Finally, Truman removed Forrestal. Reports of the secretary's increasingly nervous state reached Truman, and indeed Forrestal sometimes called the president several times and repeated previous conversations. Unhappy with Forrestal and able to pay off a political debt by giving Louis Johnson the defense post, Truman asked for Forrestal's resignation on March 1. The decision was final, yet Forrestal hesitated to prepare his letter of resignation. After Truman asked him to quit, he returned to his office and brooded over a statement. He telephoned Eberstadt (and possibly Patterson) in New York to discuss the wording of the resignation statement.[47]

Truman considered Forrestal's resignation a long overdue political and administrative decision, but for the partners in national security it became an intensely personal moment. Eberstadt's confidential message to his best friend on March 3 captured the depth and intensity of this historic relationship.

> You don't need any words, oral or written, from me to express my feelings on the type of service that you have rendered our country over these terribly difficult nine years. The annals of our country will be searched in vain to find anyone who devoted his time and talents more unselfishly, more industriously, or with a greater devotion to the sole objective of the welfare and security of our country.
>
> I cannot regard your present departure as the end of your service to the country. The lessons of history are all to the contrary. A layoff now will improve your perspective and contribute toward a clarification and firming-up of your views on national problems so that when the need arises, as indeed it will, you will perform even more valuable services in the future than you have in the past, and that's saying a whole lot.
>
> I have no idea what your plans are, but wherever I am there is always a place for you on any basis that you want it. And whatever I can do to aid in working out any plans that you have is at your disposal.[48]

Eberstadt knew that Forrestal's public service was over. But Forrestal refused to admit that he was leaving. He told Eisenhower that he could not turn the office over to the likes of Johnson, a selfish politician who would undermine all their hard work. Forrestal asked Eisenhower several times whether he should abandon Washington. "He's nervous, upset, preoccupied and unhappy," the general observed on March 19. Eisenhower and Marx Leva tried to get Forrestal to take a ten-day leave until Johnson's swearing-in as the new secretary of defense at the end of the month. Forrestal's telephone log suggested that he left Washington briefly during March to visit Monks Corner, South Carolina, possibly to see Baruch. Throughout the remainder of that last month in office, Forrestal stayed resolutely at his desk as Symington and relentless newspaper and congressional critics increased their personal attacks on him. "The situation grows intolerable," Eisenhower observed. Then at a critical moment, the general's own physical health broke and he flew to Key West to recuperate, leaving Forrestal alone. Lewis Strauss found Forrestal's predicament devastating: "He has no home life, no one to go to when things get rough." While neither separated nor divorced, long ago the secretary had isolated himself from his wife and two sons.[49]

Forrestal telephoned Eberstadt several times during the last week of March, 1949, but, according to his personal log, found his friend always out. Finally, an increasingly concerned Leva reached Eberstadt in New York and asked him to look in on Forrestal at his Washington residence. Forrestal did not want to see anybody, then relented in Eberstadt's case and agreed to see him, but only if he came alone. Eberstadt rushed to Prospect Avenue, where he found his best friend huddled in the house with the blinds drawn. "The skin of his neck and face was hanging loosely, and his loss of weight was even more noticeable," an obviously shaken Eberstadt noted. Forrestal embraced Eberstadt, clinging to him as he repeated that "they" were out to get him. "They" had tapped his telephone and hidden in his garage. Others speculated that Forrestal meant the FBI, Zionists, or investigative reporters, but he told Eberstadt that "they" referred to the "Russians." This dramatic and emotional encounter revealed the deep feelings that the two men held for one another but were rarely (if ever) displayed openly until the very end of their nearly four-decade friendship.[50]

Eberstadt told former president Hoover that Forrestal suffered from "a complete physical and perhaps nervous breakdown." After an hour of persuasion, he persuaded the reluctant former secretary of defense to leave the capital for Hobe Sound, where Forrestal's wife, along with Artemus Gates and others were vacationing at Robert Lovett's Florida estate. Eberstadt placed Forrestal on a military plane provided by Secretary Louis Johnson and then rushed to New York, where he asked one of the nation's leading psychiatrists, William Menninger, and Forrestal's good friend John Cahill to accompany him to Hobe Sound. Upon arriving in Florida and in full consultation by all present, they asked Forrestal to return to the Bethesda Naval Hospital for treatment. That night, Eberstadt and a former naval aide and loyal friend, Adm. John Gingrich, sat watch beside Forrestal's bed. It was a "personal mission in aid of our sick friend," Eberstadt explained.[51]

Forrestal's health never improved, but he was not isolated. He had access to the daily newspapers and a telephone. Truman, Louis Johnson, Sidney Souers, Robert Patterson, and others visited him in his hospital room. Apparently Eberstadt did not, although later his son found it unimaginable that his father would not visit the hospital. "Perhaps if he did not, it was for some reason that had nothing to do with their friendship, some inconvenience," Frederick Eberstadt wrote, adding that he was sure they talked on the phone. Tragedy followed. On the night of May 22, 1949, Forrestal fell (or jumped), with a bathrobe cord about his neck, to his death from an unguarded upper-story hospital window at Bethesda. Despite rumors, there was no evidence that anyone had pushed Forrestal. Three days later, Eberstadt joined the tearful, grief-stricken Patterson, Hoover, and other invited honorary pallbearers in Arlington National Cemetery. "After 30 years of close friendship," Eberstadt told New Jersey senator H. Alexander Smith, "I cannot yet realize that the awful tragedy has really happened and that my friend is gone, definitely and finally."[52]

10. The Eberstadt-Forrestal Legacy

Without Forrestal, Eberstadt's significance as an organizational facilitator and intermediary diminished rapidly. Efforts to discredit and remove him from the national security process began even as Forrestal lay in his Bethesda hospital ward during April and early May, 1949. Secretary of the Army Kenneth Royall pressed Hoover to disavow the Eberstadt task force's conclusions, while Air Force Secretary Symington urged the Senate Armed Services Committee to ignore Eberstadt's advice for more civilian control over air force budgeting and spending. "It is no secret that the private citizen who suggested the above," Symington told committee chairman Millard Tydings, "has had little actual operating experience and often makes his point consistently by talking against the military."[1]

After Forrestal's death, Eberstadt kept active in national affairs through Eisenhower. He visited and corresponded with the former general, who was at the time president of Columbia University. He quoted Eisenhower in his own talks before the World Affairs Council, New York Board of Trade, and other organizations. Eberstadt recognized that Eisenhower shared (and promoted) the same corporatist values and philosophy held by Eberstadt. For the moment, Eberstadt seemed to substitute Eisenhower for his late friend Forrestal. He treated the university president as his personal client, sharing the names from his Good Man List and arranging for "Ike" to meet the most influential New York business, radio, and political personalities. He invited Ike out to meet former president Hoover at his Target Rock estate on Lloyd Neck, Long Island.[2]

Eisenhower encouraged Eberstadt's attention. He needed this Columbia law school alumnus and member of the corporate elite to raise funds for and to promote the university's new Institute of War and Peace Studies. Eisenhower also wanted Eberstadt to help build the American Assembly, envisaged as the ideal forum for those on the Good Man List and government, business, labor, and professional leaders and managers to meet informally, study national issues, and make policy recommendations. "Those who help Columbia help America," Eisenhower explained. He expected Eberstadt to help both. Moreover, Eisenhower agreed with Eberstadt's evaluation of international affairs, particularly the emphasis on rebuilding the German and Japanese economies as cornerstones for a new world balance of power. "I believe that you and I could put down in very simple, almost diagrammatic form," he wrote Eberstadt, "the core of the central structure of our ideas concerning foreign affairs, including national security, and the relationship of these to the many great internal problems that we are now trying to solve."[3]

But Eisenhower and Eberstadt never developed a close relationship, despite their corporatist inclinations and common outlook that eventually became part of an internationalist perspective in the Republican Party. There were none of what might be called the deeper Good Man bonds that had existed between Eberstadt, Forrestal, and many of those whose names were on the Good Man List. Though Eisenhower welcomed Eberstadt's role in linking him to the likes of former president Hoover, the Luces and Dulleses, and Dewey's campaign manager Herb Brownell, he had no wish to cultivate the personal ties with Eberstadt that marked inclusion on the Good Man List. Eisenhower never considered Eberstadt for an official post or, after the presidential election of 1952, sought his advice and counsel on national security matters. Probably, Eisenhower resented Eberstadt's continued hostility toward the army's version of defense unification and remembered his role in blocking and emasculating General Marshall's single-department plan.[4]

Eberstadt's many newspaper friends offered a better, more reliable forum than Eisenhower to project and defend a putative Forrestal-Eberstadt national security legacy. Between Forrestal's death and his own in 1969, Eberstadt self-consciously promoted such a legacy of corporatist structures for American national security organization through Arthur Krock, Henry Luce, Hanson Baldwin, and other press allies. They had to educate the American public, he reminded Krock. When Truman abandoned what Eberstadt considered the vital National Security Resources Board mechanism, Eberstadt told Krock that the president had bankrupted the one agency capable of mobilizing the national economy for total war. "No plan – not even an adequate emergency program – has yet been approved and put on the shelf for use if and when," he explained. Whenever someone tried to alter or dismantle basic elements of their national security structure through defense reorganization, Eberstadt fought back through his newspaper associates. He particularly countered accusations that the loosely knit system had caused Forrestal's untimely death. "I never felt that this was the major cause of Jim's troubles," Eberstadt informed Krock. Luce promoted Eberstadt's ideas, once devoting a full page in *Time* magazine to his friend's internationalist speech, "A Balance of Power for Peace," before the World Affairs Council.[5]

Friends wanted Eberstadt to preserve and publicize their version of national security organization and philosophy by writing a Forrestal biography or by editing Forrestal's "heavily sealed" diaries, widely rumored to contain many dark secrets. Barely a month after Forrestal's death, Houghton Mifflin president Henry Laughlin asked Eberstadt to undertake such a project. Laughlin reminded Eberstadt that he had known Forrestal "intimately for 40 years" and understood that enigmatic figure better than anyone else. "I think the ideal person to edit Jim's diaries and make the best possible book is yourself," Laughlin explained. Eberstadt was not interested. He recommended instead that journalists Hanson Baldwin, Eugene Duffield, and Walter Millis edit the Forrestal diaries. In fact, Millis and Duffield produced an edited version in 1951 that preserved intact the Forrestal-Eberstadt approach to national security (the diary was actually a tract on Forrestal's national security views). No one attempted a Forrestal biography until 1962, when Eberstadt's wartime social science advisor, Robert Howe Connery, and naval historian Robert G. Albion, both employed in the Navy Department during the war, prepared a study of Forrestal's tenure as undersecretary and secretary of the Navy. The following year, political scientist Arnold Rogow published a sparsely documented psychobiography that focused on Forrestal's personality, behavior, and mental illness—a book that Eberstadt found unsatisfactory and felt had badly distorted its subject. "It seems that the author in a number of instances did not hesitate to sacrifice accuracy to interest," Eberstadt commented to Leva.[6]

Eberstadt never wrote anything about his best friend to correct Rogow's interpretation. Forrestal's death deeply affected Eberstadt, and he could not trivialize or explain in a biography for public consumption an intimate relationship based on the innermost personal, emotional, and socio-intellectual attachment. There was something protective, almost defensive, about their relationship. Their extensive correspondence developed their public views and activities but revealed no more about Forrestal than that uncovered by Rogow, Albion, or Connery. Possibly Eberstadt and Forrestal never discussed intimate matters or expressed their private feelings with each other. In any case, Forrestal displayed little about his own character or attitudes toward personal matters in his communication with his best friend, thus continuing to frustrate prospective Forrestal biographers. Perhaps Eberstadt would have obscured this private side even if he had undertaken to write a Forrestal biography.

Legal complications that tied up Forrestal's diaries and papers also delayed any biographical work. The White House and the Navy and Defense departments, family, and estate executors Clarence Dillon and the New York Trust Company had to screen and clear material before anyone could prepare a comprehensive study of the former defense official. Eberstadt wrote Dillon that his activities in this matter depended upon the Forrestal family's wishes. Apparently the immediate family never asked for Eberstadt's help in compiling a personal biography. From time to time, Eberstadt considered writing his own memoirs or recollections of the wartime mobilization experience, much as his dear friend Baruch had after the first world war. Eberstadt accumulated and saved boxes of correspondence and papers from his public service, particularly as it related to Forrestal. In the end, he told WPB assistant John Fennelly that he had never had the energy or the "guts" to undertake a major writing task. Finally, Eberstadt joined former general Lucius Clay in financing Fennelly's *Memoirs of a Bureaucrat,* hoping that his thin volume might serve, as Clarkson's study of the War Industries Board had in 1923, as a promotional tract for their organizational ideals.[7]

Eberstadt continued to testify before congressional committees or presidential commissions that periodically between 1953 and 1969 investigated the defense establishment and recommended further reorganization. For Eberstadt, efforts to consolidate the defense organization continued a long struggle between the New Dealers and corporate organizational experts begun during the early debate over mobilization policy and forms. He felt that reorganization threatened the Forrestal-Eberstadt work because it led to further concentration of the state's authority. His one-time role as an organizational innovator now became that of an opponent to further national security reorganization. The structures created by the National Security Act of 1947 and amended in 1949, Eberstadt informed the Rockefeller Committee in 1953, worked well when properly managed and administered by organizational experts. He explained that any problems in the system arose from the lack of a consistent American foreign, economic, and national defense policy. "The National Security Organization was planned to function as a coordinated whole," he explained. "It has not done so." Individual elements remained unconnected. Eberstadt lamented that the National Security Council had not fulfilled its intended mission, and he termed the National Security Resources Board a "feeble appendage."[8]

Eberstadt participated occasionally in national security affairs, but without Forrestal's patronage he remained an outsider. He worried that Eisenhower's defense reorganization efforts in 1953 and again in 1958 sought to resurrect the Marshall single-department plan of 1945 as the basis for national security consolidation. Thus, Eberstadt mobilized former allies from earlier unification battles to fight against the "hurtful changes" that redirected national security organization along the militaristic "army pattern." The "ghost of an overall Chief of Staff" had arisen again, Eberstadt cabled Adm. Arthur W. Radford. He asked Hoover to join the struggle against big government organization for defense. "So far as the Secretary's authority is concerned, I believe that the amendment of 1949 conferred upon the Secretary as much authority as the English language can convey," Eberstadt wrote the former president. He reminded Hoover that their task force had heard from hundreds of witnesses and prepared volumes of analyses before making recommendations, while the Rockefeller Committee had interviewed twenty-two people and then prepared a brief paper.[9]

In the early 1950s an expert in the Defense Production Administration who carried on the work of the corporatists by translating the Controlled Materials Plan into a mobilization readiness program during the Korean War thought that Eberstadt had preserved the corporatist national security system through his intervention. "The initiative that you displayed in talking to some of your friends did a lot to augment and marshall the opposition" to the army's design for further consolidation, he wrote Eberstadt. Several years later J. D. Hittle, a Marine Corps officer who headed a so-called "truth squad" to lobby against defense reorganization, agreed that Eberstadt's continued presence helped protect individual service interests. "Your positions and thoughts have been a beacon light for our efforts ever since this issue arose," Hittle assured the aging national security expert.[10]

Eberstadt's "beacon light" was fading. He had not obtained Hoover's support for the fight against Eisenhower's defense reorganization, and he gradually drifted away from the former chief executive's inner circle of citizen advisors. He declined Hoover's invitation to join America's business and public elite at the Bohemian Grove retreat in California. Finally, Eberstadt resigned from the Hoover Commission. "I would hate to see the Hoover Commission prestige employed to advocate matters which were not within the frame of reference of the commission," he explained to the former president. Simultaneously, Eberstadt severed connections with the American Assembly that he had helped inaugurate as a bastion for those organizational and managerial experts who Eberstadt included on his Good Man List but now promulgated a big-government philosophy. The American Assembly suggested that Forrestal and Eberstadt had designed the National Security Act of 1947 to avoid real national defense unification. This idea angered Eberstadt.[11]

His advancing age and problems over control of his investment company taxed Eberstadt. "I know you will be pleased to know that I am getting older, crankier, fatter, and balder," he wrote Forrestal's widow, Josephine Ogden, in 1959, adding that he had "degenerated into a less agreeable companion than ever." When former Forrestal aide Marx Leva, a member of President-elect John Kennedy's ad hoc commission on defense, recommended Eberstadt in 1960 for the post of secretary of defense in the incoming administration, Eberstadt replied that he ought to sue his old comrade for trying to put him in a job that would no doubt kill him. Though outwardly vigorous and active, apparently Eberstadt suffered from nervous exhaustion somewhat akin to Forrestal's final illness. Eberstadt seemed more interested in fishing, spending time with his grandchildren, or visiting his

Lyford Cay retreat in the Bahamas than in further public service. He was over seventy, he reminded Leva.[12]

Eberstadt felt that America had failed to learn from past experiences or from the teachings of the two generations of organizational experts that had adorned the Good Man List. Many of these had died during the past few years, including Robert Patterson, J. F. Dulles, H. B. Swope, and Paul Shields. Few new names replaced them. Eberstadt lamented that no one recognized that the coordinating and balancing mechanisms in the Dawes Plan or Controlled Materials Plan could solve the inflationary, budgetary, or foreign trade dilemmas of the 1960s. Even parts of the national security organization he had designed had changed so much as to be almost unrecognizable to Eberstadt. "The main difficulty with the Defense Department as it exists today," he told former president Hoover, "is that the Secretary's office is so top-heavy and over-stuffed with personnel that he has not time to deal with those things that really concern him."[13]

In Hanson Baldwin's sympathetic estimate, what remained of a putative Eberstadt-Forrestal national security legacy died with Eberstadt on November 11, 1969. "Curiously, Eberstadt fashioned the National Security Act of 1947 to prevent too much centralization, which Forrestal, the Navy, Eberstadt, etc. opposed," Baldwin observed in 1970. "Yet it has come full circle now," he wrote, "to maximum centralization – first in the Defense Department, now in the White House." As Baldwin suggested, each reorganization had further distorted the original national security structure. The Office of the Secretary of Defense grew ever larger and more powerful. JCS planning became further isolated from civilian policymaking or budgeting, boards failed to mobilize resources for the Korean War and were dropped, and individual services lost effectiveness in the name of unity. Instead of economy, efficiency, and security, the national security organization had bred bitter interservice rivalries, bloated bureaucracy, layered authority, and near paralysis in planning. It was, critics claimed, a badly flawed organization.[14]

Eberstadt and Forrestal knew from the outset that they had built a flawed national security organization. They expected it to undergo continual adjustment and orchestrated the first great reorganization in 1949 that created the Department of Defense (DOD), eliminated the service secretaries from the Cabinet, and established a uniform budgetary and accounting structure. Eberstadt explained in 1961 that the act contained built-in flaws. "Far be it for me to undertake to defend any of these; but I have been associated with these matters long enough to know that it will be a happy day, indeed, when we have a National Security Organization, and more particularly a military setup, which is at once adequate and efficient and free from waste, rivalries, boards and committees," he told the New York Chamber of Commerce. "I would be inclined to go so far as to say," he added, "that the military organization which is free from these ailments is unlikely to be either effective or efficient."[15]

Eberstadt and Forrestal had, despite the compromise, achieved their basic objective in designing and building a national security institution. They had created permanent machinery for war planning, economic mobilization, and military operations. They had intended all along, H. Struve Hensel observed, "to lay the organizational foundation which will best preserve our national security *for all time*." Eberstadt believed that they had succeeded. A few months before his death in 1969, he ruminated about his national security legacy: "I think the general pattern of an organization in our government geared to the realities of international life and the pressures of international life . . . may prove to have been my most important contribution."[16]

Eberstadt's general pattern of organization endured basically unaltered through the 1980s. Reorganization added new structures and dropped old ones, but the two-tiered system of national security organization and national military establishment remained unchanged. The National Security Council continued over the decades as the heart of this system, although different presidents used it in different ways. Truman neglected the council until the Soviets tested an atomic bomb and North Korea invaded South Korea. He then began to use the NSC as the type of policy forum envisaged by Eberstadt and Forrestal, seeking its guidance on a major restatement of administration policy prepared by State Department planners Dean Acheson and Paul Nitze, a former Dillon, Read vice-president and Forrestal advisor. The document emerged from the council as NSC-68. It echoed Forrestal's incessant warnings during his tenure as defense secretary that U.S. military strength remained inadequate to meet America's international commitments. NSC-68 recommended lifting Truman's ceiling on defense budgets and expanding American economic and military strength in order to take the offensive against Soviet communism. NSC-68 related defense spending, as Eberstadt and Forrestal had, to larger political-economic development.[17]

Possibly Nitze or the NSC staff consulted Eberstadt during the study because he revealed in a letter to Krock in the early summer of 1950 that he knew about the secret NSC-68 paper. Eberstadt told Krock that he hoped that the president would now restrict consumer installment credit, curb building programs, and reduce national and local public works projects in order to create a spending gap that could be filled by a new, expanded defense program. "As the burden rises," Eberstadt continued, "controls could be extended and tightened, particularly if fortified by appropriate tax legislation."[18]

The NSC evolved closer to an Eberstadt-Forrestal view of the agency when President Eisenhower hired Robert Cutler as his special assistant on national security affairs. A Harvard man, organizational expert, corporate lawyer and banker, Cutler fit the mold of those found on Eberstadt's Good Man List. Moreover, Cutler developed the NSC into that type of quasi-Cabinet body that the leading scholar of corporatism, Philippe Schmitter, called an essential structure for corporatization. Eisenhower employed the NSC as a consensus-building and intermediary agency that connected the Bureau of the Budget, Defense, State, and other departments. The NSC under Eisenhower became, in the words of defense expert Douglas Kinnard, "an ideal forum in which to achieve consensus and coordination and to give an impetus to the implementation of decisions."[19]

John Kennedy dismantled Eisenhower's NSC machine, but during the height of the Vietnam War Lyndon Johnson revived it as a more formal advisory council. Johnson also employed a Senior Informal Advisory Group on Vietnam, which included some of the earlier generation of Good Men such as McCloy and the younger Dillon. Richard Nixon's national security advisor, Henry Kissinger, created an NSC empire with a large staff and elaborate interagency network that linked State, Defense, CIA, and the Council of Economic Advisors. Eberstadt and Forrestal would have recognized and been encouraged by Kissinger's linkage, through the national security organization, of economic, foreign, and military policy. The architects of the 1947 act also would have understood the Ronald Reagan administration's subsequent employment of the CIA-NSC apparatus to bypass Constitution, Congress, and the public in developing its Iran-Contra initiative. Corporatist structures, Schmitter observed, deliberately narrowed the relevant publics in the initiation and conduct of national policy.[20]

The Central Intelligence Agency continued to evolve along the course set by Eberstadt and Forrestal. Forrestal initiated expansion when he used the CIA-NSC tandem, with Truman's silent approval, to influence the Italian election of 1948. The defense secretary helped launch other CIA covert operations as well and insisted that the 1947 law never prohibited such conduct. In the early 1950s, Eberstadt's fishing companion and Lloyd Neck neighbor, Allen Dulles, became director of the CIA. He brought to the CIA the putative Good Man concepts. Dulles expanded upon the Office of Strategic Services recruitment patterns initiated by William Donovan, yet another name from Eberstadt's Good Man List. Dulles made the CIA a bastion of the Good Man organizational world by contacts with the Sullivan and Cromwell law firm, Schroder banking networks, and the Council of Foreign Relations. These elaborate socioeconomic connections, some scholars suggest, influenced CIA intervention in Iran, Guatemala, Indonesia, and Cuba during Dulles's tenure as CIA director.[21]

The NSRB, the third arm (along with the NSC and CIA) of the national security triad, faltered during the Korean War. When the Reorganization Act of 1953 abolished the board, this vital corporatist agency had seemingly been removed from the national security system. In fact, industrial mobilization planning, stockpiling, contracting, and research and development functions shifted to the defense establishment. Assistant defense secretaries and a collection of functional defense agencies replaced the NSRB, Munitions Board, and Research and Development Board nexus. Defense agencies appeared under headings of Atomic Support, Contract Audit, Communications, Intelligence, and Supply. Each contained extensive intermediary arrangements for private-public consultation and collaboration. Assistant defense secretaries became the executive vice-presidents of the giant Department of Defense (DOD) "corporation," and along with the service secretaries, made up the decentralized corporate board of directors that earlier Eberstadt found in the ideal General Motors organization.[22]

DOD historian C. W. Borklund concluded that by the early 1960s the entire department had evolved into a "user-producer" establishment that paralleled that system developed in the Navy Department during the second world war by Forrestal and his civilian administrators. This organization placed the military user on one side and, as some critics claim, badly divorced it from the civilian producer on the other side. Theoretically, the military corporate board of directors embodied in the Joint Chiefs of Staff developed an integrated strategic plan and estimated the national resources required to implement that plan. Then, civilian defense administrators translated these demands to Congress and the president. Once Congress appropriated funds, the DOD further translated these demands to research and development agencies, defense contractors, and industrial producers. The system moved closer still to the one envisaged by Eberstadt and Forrestal when Secretary of Defense McNamara took over Eberstadt's "offspring" and instituted a planning-programming-budgeting system. This system owed substantial inspiration to Eberstadt's CMP, unification report, and the NSRB Plans and Programs Division.[23]

Military spending would, in the view of Eberstadt and Forrestal, stimulate, consolidate, or restrict key economic interests. Defense contracting weeded out and eliminated smaller, weaker, and fragmented interests while promoting technologically innovative and managerially efficient large-scale corporations. The latter dominated their particular functional sector. Predictably, those innovative corporations that the partners had represented in the 1920s–including Chrysler, RCA, General Electric, Standard Oil, and Goodyear–emerged in the postwar era as some of the largest defense contractors. The architects of the 1947 national security organization created mechanisms that made it possible for the one hundred largest American corporations to gain access by the 1960s to sixty-seven percent of the defense contracts.[24]

The entire national security operation depended upon the federal budget and congressional appropriations. The size of the defense budget, Eberstadt and Forrestal insisted, related to the ability to educate Congress and public to the importance of national security. During the war, crisis generated its own argument for the mobilization of economic and military resources, and promoted an awareness of national security. After the war, the sense of crisis diminished and with it public and political interest in defense appropriations or national security machinery. Only a continual atmosphere of crisis could prevent such a recurrence. "Mr. Eberstadt said that the country was always run by crises," noted James Boyd, Munitions Board assistant, adding that "if one was not evident it had to be created to get things done."[25]

Managing perpetual crisis and a national security organization "for all time" fell naturally to the Good Men. The organizational elite and technical and managerial experts found on Eberstadt's lists became more fundamental to understanding the Eberstadt-Forrestal national security legacy than any theory, process, or structure. "Charts propose and men dispose," Eberstadt once explained. "I think that this business of getting a list of good men, keeping it classified and current," he told Marx Leva in 1947, "is just as important as stockpiling materials" or in developing any other aspect of national security.[26]

Tough, public-spirited corporate experts—men like Eberstadt, Forrestal, and those on the Good Man List—had to run national security organization. Good Men could educate public and Congress to crisis and national security. They could manage the intricate corporatist mechanisms and overcome the flaws in organization. Thus, Eberstadt and Forrestal built avenues for constant access to government and paths to move back and forth between the public and private sectors. They created national security organization in their own image and in that of their "Good Man" corporate world.

Appendix: Names from the Good Man Lists, 1929–1950

Thomas R. Armstrong, Standard Oil executive, Texas rancher
War Production Board (WPB)
Bernard Baruch, stock speculator, mining executive
War Industries Board (WIB)
David K. Bruce, investment banker, aviation executive
Office of Strategic Services (OSS)
John T. Cahill, lawyer, RCA and NBC board of directors
William Chamberlain, United Light and Power executive
Lucius Clay, U.S. Army officer, business executive
Army Service Force (ASF)
Thomas Corcoran, lawyer
New Deal advisor
Joseph P. Cotton, lawyer
World War I committees
Paul D. Cravath, lawyer
World War I committees
Donald D. Davis, General Mills executive
War Production Board (WPB)
Clarence Dillon, investment banker
War Industries Board (WIB)
William J. Donovan, lawyer
Office of Strategic Services (OSS)
Lewis W. Douglas, Arizona banker, mining and chemical company executive
World War II Boards
Allen W. Dulles, lawyer, banker
Office of Strategic Services (OSS)
John Foster Dulles, lawyer, banker, chemical company executive
State Department
Frank M. Folsom, Montgomery Ward, RCA executive, NBC board of directors
War Production Board (WPB)
George S. Franklin, lawyer

Artemus L. Gates, banker, *Time,* Inc., and aviation executive
  Navy Department
Stanton Griffis, investment banker, film company executive
  Office of War Information (OWI)
  State Department
John M. Hancock, investment banker, mining company executive
  War Industries Board (WIB)
  World War II boards
W. Averell Harriman, banker and aviation, mining, and oil company executive
  World War II boards
Dannie N. Heineman, General Electric
  World War I Committees
H. Struve Hensel, lawyer
  Navy Department
Herbert C. Hoover, mining company executive
  President, United States
  World War I Committees
Courtney Johnson, Studebaker, General Motors executive
  War Production Board (WPB)
David E. Lilienthal, public utilities administrator
  Tennessee Valley Authority (TVA)
Walter Lippmann, journalist, author
  Military Intelligence Division (MID)
Robert A. Lovett, banker
  War Department
Henry R. Luce, *Time,* Inc., *Fortune* publisher
William G. McAdoo, lawyer, public utilities administrator
  Secretary of the Treasury
John J. McCloy, lawyer
  War Department
Dean Mathey, investment banker
Paul M. Mazur, investment banker and RCA, RKO executive
Robert Moses, urban planner, public administrator
Floyd B. Odlum, United Fruit, aviation, radio company executive
  War Production Board (WPB)
Robert P. Patterson, lawyer
  War Department
David Sarnoff, RCA, NBC executive
  World War II Committees
Fred Searls, mining company executive
  War Production Board (WPB)
Paul V. Shields, investment banker, aviation company executive

Lewis L. Strauss, investment banker, RCA executive
World War I committees
Navy Department
Herbert B. Swope, New York *World* publisher, RCA, CBS, RKO, utility company executive
War Industries Board (WIB)
Paul M. Warburg, investment banker, chemical company executive
Harrison Williams, utilities, American Gas & Electric
Wendell L. Willkie, utilities, lawyer
Owen D. Young, General Electric, RCA, RKO executive

Notes

PREFACE

1. Alonzo J. Walter, Jr., "Commentary on J. L. Bryon's 'Reorganization of the U.S. Armed Forces,'" *Naval Institute Proceedings* 109 (Mar., 1983): 26; Arthur T. Hadley, *The Straw Giant: America's Armed Forces: Triumphs and Failures.*

CHAPTER 1

1. Eberstadt to McNamara, Dec. 11, 1967, Ferdinand Eberstadt Papers, Seeley Mudd Manuscript Library, Princeton University (hereafter cited as Eberstadt Papers).

2. For inside-outside phenomenon, see Thomas J. McCormick, "Drift or Mastery?" p. 322; on creative individuals in institution building, see Galambos, "Technology, Political Economy, and Professionalization," p. 493.

3. Eberstadt to Forrestal, Aug. 12, 1946, James V. Forrestal Papers (hereafter, Forrestal Papers); Connor Memorandum to Krock, p. 6, Arthur Krock Papers (hereafter Krock Papers), both in Seeley Mudd Manuscript Library, Princeton University.

4. For the difficulties faced in defining and applying corporatism as an analytical construct, see both Phillipe C. Schmitter, "Still the Century of Corporatism?" pp. 85–131, and Ronald C. Newton, "Natural Corporatism and the Passing of Populism in Spanish America," pp. 34–51, both in Frederick B. Pike and Thomas Stritch, eds., *New Corporatism;* Schmitter and Lehmbruch, eds., *Trends toward Corporatist Intermediation.*

For "New American State," see Skowronek, *Building a New American State,* and Galambos, ed., *New American State.*

5. Robert H. Salisbury, "Why No Corporatism in America?" in Schmitter and Lehmbruch, *Trends toward Corporatist Intermediation,* pp. 213–30, argues that corporatism never developed in the United States. However, Ellis W. Hawley, Gerhard Lehmbruch, Robert C. Collins, Michael Hogan, and others have demonstrated convincingly that an American "corporate liberal" variation of cor-

poratism not only developed but also played a role in the organizational revolution in the United States between 1900 and 1960. See Hawley, "Discovery and Study of a 'Corporate Liberalism,'" pp. 311–12, and Hawley, "Herbert Hoover and American Corporatism, 1929–1933," pp. 101–19, in Fausold, ed., *Hoover Presidency;* David W. Eakins, "The Origins of Corporate Liberal Policy Research, 1916–1922," in Israel, ed., *Building the Organizational Society,* pp. 163–79; Collins, *Business Response to Keynes;* Hogan, *Marshall Plan;* Lehmbruch's articles, "Consociational Democracy, Class Conflict, and the New Corporatism," pp. 53–61, and "Liberal Corporatism and Party Government," pp. 147–83, both in Schmitter and Lehmbruch, *Trends toward Corporatist Intermediation;* and Schmitter, "Democratic Theory and Neocorporatist Practice," *Social Research* 50 (Winter, 1983): 885–928.

6. Alfred D. Chandler, Jr., and Louis Galambos, "The Development of Large-Scale Economic Organizations in Modern America," in Perkins, ed., *Men and Organizations,* pp. 188–201; Chandler, "The Large Industrial Corporation and the Making of the Modern American Economy," in Ambrose, ed., *Institutions in Modern America,* pp. 92–98; Chandler, *Visible Hand;* McQuaid, "Corporate Liberalism," pp. 342–68; Cuff, "The Cooperative Impulse and War: The Origins of the Council of National Defense and Advisory Commission," in Israel, ed., *Organizational Society,* pp. 233–46; Cuff, "Bernard Baruch," pp. 115–33; Cuff, *War Industries Board;* Werking, *Master Architects;* Clarkson, *Industrial America in the World War.*

7. Collins, "Positive Business Responses to the New Deal," pp. 369–91; Collins, *Business Response to Keynes;* Hogan, *Informal Entente;* McQuaid, "The Business Advisory Council of the Department of Commerce, 1933–1961: A Study in Corporate/Government Relations," in Uselding, ed., *Research in Economic History* 1:171–97; Himmelberg, *Origins of the National Recovery Administration.*

8. McQuaid, "Corporate Liberalism," pp. 363–64; Eberstadt to Henry Luce, Dec. 6, 1942, Eberstadt Papers; Millis and Duffield, eds., *Forrestal Diaries,* p. 134 (hereafter, *Forrestal Diaries*).

9. Eberstadt's Remarks before Presbyterian Men's Club Westfield, N.J., Apr. 13, 1944, typescript; Eberstadt's airline syndicate, in Eberstadt to Leva, Mar. 18, 1947, both in Eberstadt Papers; on national radio corporation, Rogow, *James Forrestal,* p. 334.

10. *Forrestal Diaries,* p. 252; on business, labor, and government partnership, Eberstadt to Leslie Gould (financial editor, New York *Journal American*), Jan. 3, 1943, Eberstadt Papers.

11. Rogow, *James Forrestal,* p. 255; Special Group–Syndicate List: Walter Lippmann, W. A. Harriman, David Bruce, William Donovan, etc. (Good Man List, Mar. 14, 1933, Eberstadt Papers); also, see the interpretation of these men as part of a monolithic power elite, in C. Wright Mills, *The Power Elite* (New York: Oxford University Press, 1956); Domhoff, *Who Rules America?;* Barnet, *Roots of War;*

and for the most balanced approach, Burch, Jr., *Elites in American History,* 3 vols.

Lippmann, *Drift and Mastery,* probably most influenced Eberstadt and Forrestal.

12. *Forrestal Diaries,* p. 142; on techno-experts, see Eberstadt to Forrestal, Apr. 29, 1946; Eberstadt Memorandum of Thoughts, July 12, 1945; also, see Good Man lists, 1929–1947, all in Eberstadt Papers.

13. Good Man composite based on biographical information in Burch, Jr., *Elites in American History; New York Times* obituaries; *Current Biography; Who's Who;* and *Who Was Who in America.*

For a recent interpretation of this elite that perpetuates the white Anglo-Saxon Protestant establishment typology, see Isaacson and Thomas, *Wise Men.*

14. Eberstadt to Hoover, Aug. 24, 1956, Eberstadt Papers. The best "corporatist biography" is Schwarz's study of Bernard Baruch, *The Speculator;* also, Hawley, "Hoover," in Fausold, ed., *Hoover Presidency;* Joan Hoff Wilson, *Herbert Hoover: Forgotten Progressive* (Boston: Little, Brown, 1974); corporatist biographical sketches in Hogan, *Informal Entente* and *Marshall Plan;* Collins, *Response to Keynes;* and McQuaid, "Young, Swope," and "Corporate Liberalism."

15. See Hogan, *Marshall Plan.*

16. Eberstadt to Taft, Dec. 1, 1947, Eberstadt Papers.

17. See Werking, *Master Architects;* Theodore Ropp, "The Rise of American Military Power," in Ambrose, ed., *Institutions in Modern America,* pp. 107–10; Kaufman, *Efficiency and Expansion;* Skowronek, *New American State;* Morison, *Stimson.*

18. Willoughby, *Government Organization,* p. 357; Cuff, "Cooperative Impulse and War," in Israel, ed., *Building the Organizational Society,* pp. 233–46; Chambers, *To Raise an Army.*

19. Cuff, *War Industries Board;* Cuff, "Bernard Baruch," pp. 115–33; Cuff, "Herbert Hoover, Ideology of Voluntarism," pp. 358–72.

20. Willoughby, *Government Organization;* Berman, *Office of Management and Budget;* Morrow Commission, in *Forrestal Diaries,* pp. 46–47.

21. Yoshpe, "Bernard M. Baruch"; Cuff, "Bernard Baruch"; Schwarz, *Speculator,* p. 211.

CHAPTER 2

1. Mulder, *Woodrow Wilson,* pp. 159–82; Cronon, ed., *Political Thought of Woodrow Wilson,* pp. 109–41; Link, ed., *Wilson Papers,* 19:525–27.

2. Eberstadt's Newark Academy Graduation Address, June 15, 1952, Eberstadt Papers; Frederick Eberstadt, interview with author, Sept. 22, 1988; *Princeton Alumni Weekly,* Jan. 27, 1970, p. 19; see also Perez and Willett, *Will to Win,* pp. 17–19.

3. Eberstadt to Eliot, Aug. 7, 1947, Eberstadt Papers; Rogow, *James Forrestal,* pp. 49–65.

4. Albion and Connery, *Forrestal and the Navy,* p. 9; Krock Private Memorandum, "An Interview with Ferdinand Eberstadt," May 6, 1949, Krock Papers.

5. Frederick Eberstadt interview, Sept. 22, 1988; Frederick Eberstadt to the author, Sept. 22, 1988; Schneider to Christman, Jan. 11, 1971, Calvin L. Christman Private Papers, Cedar Valley College, Lancaster, Tex. (hereafter, Christman Papers).

6. Eberstadt's Notes for Bryn Mawr-Princeton Speech, n.d., Eberstadt Papers; Kennan, *Memoirs,* p. 13. Eberstadt to Eliot, Aug. 7, 1947, Eberstadt Papers.

7. Connor, "Impressions of the Secretary," Oct. 23, 1961, pp. 16–19, Krock Papers; Osgood, *Lights in Nassau Hall.*

8. Wilson's 1911 Inaugural Address, in Cronon, ed., *Political Thought of Woodrow Wilson,* p. 156.

9. Link, ed., *Wilson Papers,* 23:635–36; Eberstadt's Radio Address, transcript for ABC Headline Edition, Dec. 16, 1948, WNEW, Dec. 17, 1948, Eberstadt Papers.

10. Eberstadt to Gould, Jan. 9, 1943, Eberstadt Papers; *Daily Princetonian,* Mar. 13, 1912, p. 2, Dec. 6, 1948, p. 1; Eberstadt to Chamberlain, Sept. 25, 1950, Eberstadt Papers; *Twenty-Fifth Year Record of the Class of 1913,* Princeton University, p. 56.

11. Fox, *Discovery of Abundance,* p. 23; Herbst, *German Historical School,* p. 149.

12. Fetter, "Subsidizing of Private Charities," p. 384, and "State Supervision," p. 457.

13. Redfield, *New Industrial Day,* p. 47; see also Kaufman, *Efficiency and Expansion,* pp. 69–71; Werking, *Master Architects,* pp. 241–43; Burch, Jr., *Elites in American History,* 2:211, 235.

14. Eberstadt's undergraduate paper, "The New Industrial Day," for Fetter's Division, Mar. 30, 1912, Eberstadt Papers; Redfield, *New Industrial Day,* p. 44.

15. Bowen, *German Theories of the Corporative State;* Herbst, *German Historical School,* pp. 163–64.

16. Bowen, *German Theories of the Corporative State,* pp. 214–16; Perez and Willett, *Will to Win,* p. 22; Frederick Eberstadt interview, Sept. 22, 1988; admiration for Brentano and Schmoller, Eberstadt to his mother, Nov. 23 and 29, 1913, Eberstadt Papers; Eberstadt's personal copy of Ludwig Bernhard's *Das Polnische Gemeinwesen im Preussischen Staat,* found in ruins of Eberstadt's personal library, Target Rock mansion (now in Seeley Mudd Manuscript Library, Princeton University).

17. Lash, *Frankfurter Diaries,* p. 166; Blum, ed., *Price of Vision: The Diary of Henry A. Wallace* (hereafter, *Wallace Diary*), p. 140; Fetter to Eberstadt, Dec. 2, 1913, Eberstadt Papers.

18. *Ten Year Record of the Class of 1913,* Princeton University, pp. 59–60; Christ-

man, "Ferdinand Eberstadt and Economic Mobilization for War, 1941–1943," (Ph.D. diss., Ohio State University, 1971), p. 7.

19. *New York Times,* Jan. 2, 1923, p. 23; Eberstadt to Eliot, Aug. 7, 1947, Eberstadt Papers; Rogow, *James Forrestal,* p. 65ff; Kahn, Jr., *World of Swope;* Krock, *Memoirs.*

20. Rogow, *James Forrestal,* pp. 67–68; Frank P. Leslie, *James Forrestal: Princeton Class of 1915* (pamphlet, privately printed, 1951).

21. Wood quoted in Weigley, *Towards an American Army,* p. 216; Millis, ed., *American Military Thought,* p. 274; Lane, *Armed Progressive.*

22. Ferdinand Eberstadt, interview with Christman, July 18, 1969 (transcript in Seeley Mudd Manuscript Library, Princeton University).

23. Swaine, *Cravath Firm,* 3:100–101; Eberstadt to Patterson, May 18, 1943, Eberstadt Papers.

24. Leslie, *James Forrestal;* envy of war wounds in Forrestal to Maas, Aug. 11, 1945, Forrestal Papers; Gould to Krock, July 27, 1947, Krock Papers; Roosevelt, "Attitude of the Navy toward a United Air Service, June 14, 1919," File 449, General Board Records, Naval Historical Division, Operational Archives, Washington, D.C. (hereafter NHD).

25. See Kahn, Jr., *World of Swope;* Krock, *Memoirs;* Rogow, *James Forrestal.*

26. Eberstadt to Francis McAdoo, July 11, 1961; W. G. McAdoo to Eberstadt, Apr. 17, 1926, both in Eberstadt Papers; Hogan, *Informal Entente,* p. 114; Otis L. Graham, Jr., "William Gibbs McAdoo," in *Dictionary of American Biography,* 15:479–82.

27. Cotton, "Recent New York Legislation," pp. 232–37; *New York Times,* Mar. 11, 1931, pp. 1, 21; Swaine, *Cravath Firm,* p. 264; and see McQuaid, "Corporate Liberalism," pp. 342–68, for concept of the liberal fringe of the corporate world.

28. *New York Times,* Mar. 11, 1931, pp. 1, 21; Hogan, *Informal Entente,* pp. 2–3; Swaine, *Cravath Firm,* p. 16.

29. Joseph Cotton and Dwight Morrow, "International Cooperation," *Atlantic* (June, 1919): 805–808; Swaine, *Cravath Firm,* p. 122.

30. Swaine, *Cravath Firm,* pp. 1–3; Mayer, *Emory Buckner;* Eberstadt's Daily Appointment Books, 1924–1932, Eberstadt Papers. The "Good Man" lawyers included John Foster and Allen Dulles, Emory Buckner, Elihu Root, Jr., Robert Patterson, William Donovan, John Lord O'Brian, John J. McCloy, and Bernhard Knollenberg. See Hogan, *Marshall Plan,* p. 10.

31. Eberstadt to Eliot, Aug. 7, 1947; Eberstadt's Daily Appointment Book for Aug., 1925, both in Eberstadt Papers.

32. Maier, *Recasting Bourgeois Europe;* Hogan, *Informal Entente,* pp. 68, 77.

33. *New York Times,* Dec. 12, 1924, p. 6.

34. Eberstadt to Chamberlain, Mar. 29, 1944; Eberstadt Appointment Book,

Dec., 1924, June, July, and Aug. entries, 1925, all in Eberstadt Papers; Meyer, *Bankers' Diplomacy,* p. 72n36.

35. E. R. Harriman to Christman, May 27, 1971, Christman Papers; Eberstadt's Daily Appointment Books, 1931–32, Eberstadt Papers; Burch, Jr., *Elites in American History,* 3:142; Janeway, *Economics of Crisis,* p. 208; Frederick Lewis Allen, *The Lords of Creation* (New York: Harper and Brothers, 1935), pp. 254, 342–43.

36. Willett Memo for Eberstadt, Dec. 7, 1927, Eberstadt Papers; Edward F. Willett, interview with Christman, Aug. 12, 1970, Christman Papers; Mayer, *Emory Buckner,* p. 238; Frederick Eberstadt interview, Sept. 22, 1988.

37. Eberstadt to Forrestal, Nov. 24, 1927; Eberstadt, "A Friend's Recollection of Melchior," in *Dr. Carl Melchior,* pp. 121–22; Eberstadt's Berlin Diary fragment, Feb. 19, 1926, Eberstadt Papers.

38. *New York Times,* Aug. 4, 1927, p. 32, Dec. 22, 1927, p. 34; Eberstadt's Daily Appointment Book, Nov. 18, 1932, Eberstadt Papers; Maier, *Recasting Bourgeois Europe,* p. 580.

39. Griffis, *Lying in State,* pp. 56–57; Willett interview with Christman, Aug. 12, 1970; Swaine, *Cravath Firm,* p. 518, and passim.

40. Cotton and Morrow, "International Cooperation," pp. 805–808.

41. Eberstadt interview, transcript, p. 5, Dulles Oral History Project, Seeley Mudd Manuscript Library, Princeton University; "Report on Acceptance Credit Granted to Swedish Match Company Stockholders by American Banks, Sept. 14, 1932," Eberstadt Papers; *New York Times,* Oct. 4, 1929, p. 44.

42. Christman, "Eberstadt," pp. 20–27; *Fortune* 19 (Apr., 1939): 72–75; E. R. Harriman to Christman, May 27, 1971, Christman Papers; Perez and Willett, *Will to Win,* p. 51.

43. Eberstadt to Forrestal, June 14, 1933, Eberstadt Papers; also, Kennedy, *Banking Crisis,* pp. 103–28.

44. Eberstadt to W. A. Harriman, Jan. 17, 20, and 21, 1934; "Lists Furnished Mr. Harriman for N.R.A. Work," Jan. 24, 1934; "This Federal Tax Idea Would Quickly Revive Drooping Business," *Printer's Ink,* Nov. 18, 1937, all in Eberstadt Papers; David W. Eakins, "Business Planners and America's Postwar Expansion," in Horowitz, ed., *Corporations and the Cold War,* pp. 143–71.

45. Eberstadt to Forrestal, Nov. 15, 1934; Eberstadt Diary, fragment, May 3, 1943; Eberstadt to Shields, Apr. 28, 1944, all in Eberstadt Papers; also, Parrish, *Securities Regulation,* p. 57 and passim; Borklund, *Men of the Pentagon,* p. 14.

46. Eberstadt Memo to self, May 29, 1943; Eberstadt to Bracken, Apr. 12, 1948, both in Eberstadt Papers.

47. Eberstadt to Krock, Nov. 28, 1944; Eberstadt to Forrestal, May 22, 1941, both in Eberstadt Papers; Allen, ed., *Democracy and Finance: The Addresses and Public Statements of William O. Douglas* (hereafter, *Douglas Addresses*), p. 1.

48. Catton, *War Lords*, p. 102; Schwarz, *Speculator*, p. 323ff. George C. Marshall, in a Nov. 28, 1939, letter to Lorenzo D. Gasser complained about lack of coherent mobilization plan (Bland, ed., *Marshall Papers*, 2:113 [hereafter, *Marshall Papers*]).

49. Eberstadt to Chamberlain, Feb. 29, 1940, Eberstadt Papers; Dewey, *Case against the New Deal*, pp. 120, 127; Chamberlain to Eberstadt, May 20, 1940, Eberstadt Papers.

50. Baruch to Watson, May 22, 1940, President's Official File 4032, FDR Library, Hyde Park, N.Y.

CHAPTER 3

1. Janeway, *Struggle for Survival*, p. 136; FDR to Forrestal, Apr. 11, 1940; Forrestal to FDR, Apr. 22, 1940, President's Personal File 6567, both in FDR Papers; Smith Memo to President, June 2, 1941, Smith Papers, FDR Library, Hyde Park, N.Y.

2. Forrestal to Early, June 28, 1940, President's Official File 4032; Forrestal to the President, May 10 and 26, 1940, President's Personal File 6567, all in FDR Papers.

3. Memorandum (by unknown author) on James Forrestal, June 26, 1940, President's Secretary's File–Navy Department–Forrestal, FDR Papers; Catton, *War Lords*, p. 23.

4. Catton, *War Lords;* Burch, Jr., *Elites in American History*, 3:69–80.

5. Eberstadt to Dillon, Sept. 27, 1942, Eberstadt Papers.

6. Eberstadt Memo to self, Feb., 1943; Eberstadt Speech to Michigan Industrial Leaders, typescript, July 3, 1942, both in Eberstadt Papers.

7. Brownlow, *Passion for Anonymity*, p. 426; Forrestal Memo to FDR, July 1, 1940, OF 4032, FDR Papers; Forrestal to Hitchcock, Aug. 3, 1940; Hitchcock to Forrestal, July 29, 1940, both in Forrestal Papers.

8. *Marshall Papers*, 2:254; Burch, Jr., *Elites in American History*, 3:74; Samuel Lobdell, "Frank Knox," in Coletta, ed., *Secretaries of the Navy*, 2:685; Stimson and Bundy, *On Active Service;* John L. Sullivan Oral History, 1974, Harry S. Truman Library, Independence, Mo. (hereafter, HSTL).

9. Forrestal to Hitchcock, Aug. 3, 1940; Forrestal to Towers, July 23, 1940, both in Forrestal Papers; Albion and Connery, *Forrestal and the Navy*, pp. 7–8.

10. Sexton to Secretary of the Navy, July 1, 1940, serial 1959, File 425, General Board Records, NHD.

11. Albion and Connery, *Forrestal and the Navy*, pp. 38–58.

12. Forrestal to Eberstadt, July 22, Aug. 22, Oct. 7, 17, 21, 1940; Eberstadt to Forrestal, Sept. 13, Oct. 8, 9, 13, 17, and Dec. 12, 13, 17, 1940, all in Eberstadt Papers; Albion and Connery, *Forrestal and the Navy*, pp. 30–36; Connery, *Navy and Industrial Mobilization*, p. 62ff.

13. Forrestal to Eberstadt, Sept. 13, 1940, Eberstadt Papers; Albion and Connery, *Forrestal and the Navy,* pp. 63–64.

14. Catton, *War Lords,* pp. 153–54; Smith Memo to the President, June 2, 1941, Harold D. Smith Papers, FDRL (hereafter, Smith Papers); *Marshall Papers,* 2:117–19.

15. Berman, *Office of Management and Budget,* pp. 11–24.

16. Smith Memo to the President, June 2, 1941, Smith Papers; Hewes, Jr., *From Root to McNamara,* pp. 316–17.

17. *Marshall Papers,* 2:523; Earle to Eliot, July 10, 1941, Edward M. Earle Papers, Manuscript Library, Princeton University (hereafter, Earle Papers); Schwarz, *Speculator,* p. 370ff.

18. Smith Memo to the President, June 2, 1941, Smith Papers.

19. Janeway, *Struggle for Survival,* p. 136.

20. Catton, *War Lords,* p. 102; Eberstadt to Forrestal, Feb. 14, 1941, Eberstadt Papers; Schwarz, *Speculator,* p. 371ff.

21. Catton, *War Lords,* pp. 4, 17, 244; Nelson, *Arsenal of Democracy,* p. 198.

22. Eberstadt to Hensel, Dec. 28, 1945, Eberstadt Papers; Nelson, *Arsenal of Democracy,* pp. 178, 193.

23. Eberstadt to Forrestal, June 18, 1941; Eberstadt to Chamberlain, July 14, 1941; Forrestal to Eberstadt, May 14, 1941, all in Eberstadt Papers.

24. Eberstadt to Forrestal, May 22, 1941, Eberstadt Papers.

25. Eberstadt to Forrestal, Feb. 14, May 22, 1941, Eberstadt Papers.

26. R. Elberton Smith, *Army and Economic Mobilization,* pp. 73–97; Christman, "Eberstadt," pp. 59–156; Connery, *Navy and Industrial Mobilization.*

For the debate over the ANMB role in the formation of a military-industrial complex, see Koistinen, "'Industrial-Military Complex,'" pp. 819–39, and Cuff, "Organizational Perspective," pp. 250–67; Pursell, *Military-Industrial Complex.*

27. Connery, *Navy and Industrial Mobilization,* p. 42; Smith, *Army and Economic Mobilization,* pp. 37–39.

28. Smith, *Army and Economic Mobilization,* p. 76; Schwarz, *Speculator,* pp. 356–58.

29. For example, see Eberstadt to Forrestal, May 22, 1941, Eberstadt Papers.

30. Eberstadt to Forrestal, Oct. 2, 1941; Eberstadt Diary, fragment, Oct. 30, 31, 1941, both in Eberstadt Papers; Nelson, *Arsenal of Democracy,* pp. 129–35.

31. *New York Times,* July 14, 1971, p. 16.

32. Dorr Memorandum for Undersecretary of War, Apr. 9, 1942; Eberstadt Diary, fragment, Oct. 31, 1941, both in Eberstadt Papers; Greenbaum, *Lawyer's Job;* Millett, *Army Service Forces,* p. 29ff.

33. Critchlow, *Brookings Institution,* pp. 9–12; Kiplinger, *Washington Is Like That,* pp. 331–32.

34. "Talk with Stacy May," Oct. 31, 1941, Eberstadt Diary, Eberstadt Papers.

35. Rutherford referred to pages 37, 48, 49, 55, 176, 180, and 301, in Clarkson,

Industrial America in the World War, noted in Eberstadt Diary, Oct. 31, 1941, Eberstadt Papers; for Rutherford's biography, see *National Cyclopaedia of American Biography* 50, p. 162.

36. Clarkson, *Industrial America in the World War*, p. 180; see also Cuff, "Bernard Baruch," pp. 115–33; Cuff, "Herbert Hoover," pp. 358–72.

37. Gulick, *Administrative Reflections*, pp. 41–43; see Cuff, "Organizational Perspective," for institutional discontinuity interpretation.

38. Clarkson, *Industrial America in the World War*, pp. 37, 55, 176, 180 (quote).

39. Ibid., pp. 300–301, 488 (quote).

40. "Talk with Baruch," Nov. 3, 1941, Eberstadt Diary, Eberstadt Papers.

41. "Talk with Admirals Pickens and Stark," Nov. 7, Oct. 31, 1941, Eberstadt Diary, Eberstadt Papers.

42. "Talk with Mr. Donald Nelson," Nov. 12, Scott in Eberstadt Notes, Oct. 31, 1941, Eberstadt Diary, Eberstadt Papers.

43. Christman, "Eberstadt," p. 59ff; Connery, *Navy and Industrial Mobilization*, pp. 158–60.

44. Gladieux Memorandum on Army and Navy Munitions Board Proposal, Jan. 1 (1942), file 39.32, Bureau of the Budget Records in Record Group 51, NA.

45. Nelson, *Arsenal of Democracy*, p. 198.

CHAPTER 4

1. David E. Lilienthal, *The Journals of David E. Lilienthal* (hereafter *Lilienthal Journals*), 1:545; Eberstadt Speech, transcript, July 3, 1942, Eberstadt Papers.

2. Eberstadt Speech, transcript, July 3, 1942, Eberstadt Papers.

3. Ferdinand Eberstadt interview with Christman, July 18, 1969, Princeton; Casey quoted in, Blum, ed., *Wallace Diary*, p. 140.

4. Schwarz, *Speculator*, p. 432; Connery, *Navy and Industrial Mobilization;* Greenbaum, *Lawyer's Job*, p. 148.

5. Ferdinand Eberstadt interview with Christman, July 18, 1969, Princeton; Supplement to General Memorandum, n.d. (Feb., 1943?), Eberstadt Papers; Millett, *Army Service Forces*, p. 216; Blum, ed., *Wallace Diary*, pp. 94, 119–20.

6. Ferdinand Eberstadt interview with Christman, July 18, 1969, Princeton.

7. Ibid.

8. Blum, ed., *Wallace Diary*, p. 121.

9. Nelson, *Arsenal of Democracy*, p. 363; Ferdinand Eberstadt interview with Christman, July 18, 1969.

10. Ferdinand Eberstadt interview with Christman, July 18, 1969, Princeton; see also Schwarz, *Speculator.*

11. H. Luce to Eberstadt, Sept. 22, 1942; Remarks by Eberstadt before Business Advisory Council, Carlton Hotel, transcript, Nov. 19, 1942, both in Eber-

stadt Papers; see McQuaid, "Business Advisory Council," in Uselding, ed., *Research in Economic History,* 1:171–97.

12. Ferdinand Eberstadt interview with Christman, July 18, 1969, Princeton.

13. Eberstadt to Fox, May 21, 1968; Eberstadt to Auxier, Nov. 8, 1942; Eberstadt to Leva, Apr. 1, 1965, all in Eberstadt Papers; Novick, Anshen, and Truppner, *Wartime Production Controls,* p. 155; Smith, *Army and Economic Mobilization,* pp. 567–69; Christman, "Eberstadt," p. 210.

14. Hitch, quoted in Christman, "Eberstadt," p. 227n58; Novick, Anshen, and Truppner, *Wartime Production Controls,* p. 160.

15. Herring to Milton, Oct. 20, 1942, series 39.32, Division of Administrative Management Government Organization Branch, Bureau of the Budget Records in Record Group 51, NA; Catton, *War Lords,* p. 311.

16. Nelson, *Arsenal of Democracy,* p. 270; Schwarz, *Speculator,* p. 433.

17. Memo to Eberstadt for National Association of Manufacturers Speech, Nov. 28, 1942, Eberstadt Papers; Nelson, *Arsenal of Democracy,* p. 270.

18. Nelson, *Arsenal of Democracy,* pp. 382, 387–88; Ferdinand Eberstadt interview with Christman, July 18, 1969, Princeton.

19. *Lilienthal Journals,* 1:574; Eberstadt to Patterson, Nov. 26, 1942, Eberstadt Papers; Blum, ed., *Wallace Diary,* p. 146.

20. Lash, ed., *Frankfurter Diary,* pp. 185–86; Reich to Christman, Oct. 29, 1970, Christman Papers; Blum, ed., *Wallace Diary,* pp. 136–37.

21. Eberstadt interview with Christman, July 18, 1969, Princeton; Eberstadt's Supplement to General Memorandum, n.d. (Feb., 1943?), Eberstadt to Gould, Jan. 3, 1943; both in Eberstadt Papers.

22. Catton, *War Lords,* p. 205; Burch, Jr., *Elites in American History,* 3:113–14; Schwarz, *Speculator,* pp. 436–38.

23. Catton, *War Lords,* p. 206; Ferdinand Eberstadt interview with Christman, July 18, 1969, Princeton; *New York Times,* Feb. 14, 1943, p. 1; Blum, ed., *Wallace Diary,* pp. 68, 133, 252, 287.

24. Nelson, *Arsenal of Democracy,* pp. 387–88; Schwarz, *Speculator,* p. 439; Catton, *War Lords,* p. 206.

25. Catton, *War Lords,* p. 206; Ferdinand Eberstadt interview with Christman, July 18, 1969, Princeton; Swope to Eberstadt, Oct. 27, 1942, Eberstadt Papers.

26. Catton, *War Lords,* p. 206.

27. Ferdinand Eberstadt interview with Christman, July 18, 1969, Princeton; Blum, ed., *Wallace Diary,* p. 191; Fennelly, *Memoirs of a Bureaucrat,* p. 68; Eberstadt to Patterson, May 18, 1943, Eberstadt Papers.

28. Eberstadt Memorandum and Supplement to Memorandum, n.d. (Feb., 1943?), Eberstadt Papers; Ferdinand Eberstadt interview with Christman, July 18, 1969, Princeton; Catton, *War Lords,* p. 48; Schwarz, *Speculator,* p. 451.

29. Memorandum and Supplement to Memorandum, n.d.; Eberstadt to Patterson, Sept. 12, 1946, both in Eberstadt Papers; Catton, *War Lords,* pp. 180–81.

30. Eberstadt Memorandum and Supplement (Feb., 1943?), Eberstadt Papers; Ferdinand Eberstadt interview with Christman, July 18, 1969, Princeton.

31. Eberstadt's Washington Memorandum, May 10, 1943; Eberstadt to Chamberlain, July 13, 1943, Feb. 28, 1944; Eberstadt to Lovett, July 30, 1943; Eberstadt Diary, May 7, 1943, all in Eberstadt Papers; Eberstadt to Baruch, May 18, 1943, Bernard M. Baruch Papers, Seeley Mudd Manuscript Library. Princeton University.

32. Nelson, *Arsenal of Democracy,* p. 385; Catton, *War Lords,* pp. 247–49.

33. Eberstadt Diary, May 3, 6, 13, and 26, June 15, 1943, Eberstadt Papers.

34. Eberstadt to Forrestal, Aug. 18, 1943; Eberstadt Diary, July 12, 1943; Forrestal to Eberstadt, Aug. 8, 1943, all in Eberstadt Papers.

35. Eberstadt's Washington Memo, May 29, 1943; Eberstadt Diary, June 5 and 15, 1943, both in Eberstadt Papers; *Lilienthal Journals,* 1:666; Baruch to Eberstadt, Jan. 28, 1944, Baruch Papers; also Catton, *War Lords,* p. 133; Schwarz, *Speculator,* pp. 444–45.

36. Eberstadt to Forrestal, Sept. 22, Oct. 4, 1943, Eberstadt Papers.

37. Eberstadt Memorandum, "Principles, Procedures and Organization of Industrial Demobilization," Nov. 6, 1943, Baruch Papers.

38. Ibid.

39. Catton, *War Lords,* p. 283; Burch, Jr., *Elites in American History,* 3:111n11; also, Schwarz, *Speculator,* pp. 441–71.

40. Schwarz, *Speculator,* p. 461; Patterson to Eberstadt, Nov. 21, 1944, Eberstadt Papers. See *Prefaces to Peace: A Symposium Consisting of the Following: One World by Wendell L. Willkie, The Problems of Lasting Peace by Herbert Hoover and Hugh Gibson, The Price of Free World Victory by Henry A. Wallace, and Blue-print for Peace by Sumner Welles* (New York: Simon and Schuster, 1943).

41. Eberstadt to Lawrence, Aug. 28, 1944; Eberstadt to Clare B. Luce, Aug. 28, 1944; Eberstadt to Forrestal, July 18, 1944, all in Eberstadt Papers.

42. Eberstadt's "Address to Class of 1913," Oct. 22, 1944; Eberstadt to H. Luce, Dec. 6, 1942, both in Eberstadt Papers; see also Jessup, ed., *Ideas of Henry Luce,* p. 115.

43. Eberstadt's "Address to Class of 1913," Eberstadt Papers.

44. Ibid.; see also Eberstadt's Washington Memos, June 14, 15, and 21, 1943, Eberstadt Papers.

45. Elsie (Mrs. George S.) Franklin to Eberstadt, Mar. 23, 1944, Eberstadt Papers; McCormick, "Drift or Mastery?"; Willoughby to Eberstadt, Apr. 27, 1944, Eberstadt Papers.

46. Eberstadt to Forrestal, Sept. 6, 1944; Chamberlain to Eberstadt, Feb. 12, 1944, both in Eberstadt Papers.

47. Eberstadt to Forrestal, Nov. 27, 1944, Eberstadt Papers.

48. Eberstadt to Krock, Nov. 28, 1944; Chamberlain to Eberstadt, June 2, 1944; Eberstadt to Chamberlain, Sept. 7, 1944; Eberstadt to Elliott V. Bell, Aug. 8, 1944; Bell to Eberstadt, Dec. 11, 1944; Harold Keller to Eberstadt, Sept. 7, 1944; Dewey Speech Notes, Jan. 2, 1945, all in Eberstadt Papers.

49. See *Prefaces to Peace;* Eberstadt to Krock, Nov. 28, 1944; Eberstadt to Chamberlain, Sept. 7, 1944, both in Eberstadt Papers.

50. Eberstadt to Forrestal, July 18, 1944, Eberstadt Papers.

CHAPTER 5

1. Catton, *War Lords,* p. 240.

2. *Forrestal Diaries,* p. 92.

3. Blaisdell to Christman, Dec. 22, 1970, Christman Papers.

4. *Marshall Papers,* 1:45–46, 348; 2:259, 562–63.

5. Ibid., 1:24, 646.

6. T. Roosevelt, quoted in Millis, ed., *American Military Thought,* pp. 265, 268.

7. *Marshall Papers,* 1:399–401.

8. Ibid., 1:313, 376, 633–34, 675.

9. Holley, Jr., *Gen. John M. Palmer,* p. 486; Hewes, *From Root to McNamara,* p. 50; U.S. Congress, House, Committee on Military Affairs, *Department of Defense, and Unification of Air Service Hearings,* 69th Cong., 1st sess., 1926, pp. 1005–10; *Marshall Papers,* 1:28, 36, 50, 593–94.

10. *Marshall Papers,* 1:313, 321ff; Palmer, *Statesmanship or War,* p. 180.

11. *Marshall Papers,* 1:621–23; 2:107–108, 113.

12. Palmer to Earle, June 10, 1943, Earle Papers; Holley, Jr., *Gen. John M. Palmer,* pp. 183–85, 543; Earle to Culbertson, May 15, 1943, Earle Papers.

13. Somervell Memo for Grunert, June 28, 1943, Marshall Research File, George C. Marshall Library, Lexington, Va.

14. Tompkins Memo for Marshall, Aug. 16, 1943; "A Single Department of National Defense," typescript, n.d., both in Marshall Research File.

15. Palmer to Earle, June 10, 1943, John M. Palmer Papers, Library of Congress; Marshall Conference with the President, Aug. 31, 1943, Smith Diary, Smith Papers, FDRL; McNarney Memo for COS, Aug. 24, 1943, Marshall Research File; Palmer to Miles, Aug. 16, 1943; Palmer to Tompkins, June 18, 1943, both in Palmer Papers.

16. King to Edson, Aug. 24, 1949, Ernest J. King Papers, Library of Congress; Knox to the President, Aug. 17, 1943, Knox-Forrestal File, Naval Historical Division (hereafter, Forrestal File, NHD).

17. Yarnell to Halsey, Aug. 4, 1943, William F. Halsey Papers, Library of Congress; Buell, *Master of Sea Power,* pp. 364–67; King to Horne, Sept. 1, 1943, King Papers.

18. Knight Endorsement on Conference of Cooperation between Army and Navy, July 6, 1915; Clark Memorandum for Chief of Staff, May 2, 1919, both in William S. Sims Papers, Library of Congress; Yarnell to Earle, Mar. 5, 1942, Sept. 30, Oct. 6, 1943; Earle to Yarnell, Mar. 4, 1942, Mar. 6, Oct. 6, 1943, all in Earle Papers; Yarnell, "Department of War," *U.S. Naval Institute Proceedings* Aug., 1943, pp. 1099–1101.

19. LeBreton to Yarnell, Aug. 16, 1943; Yarnell to Secretary of the Navy, Nov. 6, 1943; Nimitz to Yarnell, Sept. 1 and 23, 1943; Brown to Yarnell, Aug. 9, 1943; Sherman to Yarnell, Sept. 8, 1943, all in Adm. Harry E. Yarnell Papers, Naval Historical Division, (hereafter, Yarnell Papers, NHD).

20. Moss to Earle, Nov. 22, 1943, Earle Papers; Palmer to Yarnell, Sept. 2, 1943, Palmer Papers; King to Horne, Sept. 1, 1943, King Papers; Yarnell to Earle, Sept. 30, Oct. 8, 1943; Earle to Yarnell, Oct. 6, 1943; Yarnell to Culbertson, Apr. 6, 1944; Culbertson to Yarnell, May 1, 1944; Earle to Moss, Oct. 29, 1943, all in Earle Papers.

21. Palmer to Miles, Sept. 3, 1943; Palmer to Tompkins, Sept. 4, Oct. 4, 1943, all in Palmer Papers.

22. Tompkins Memo for COS, Oct. 11, 1943; Tompkins to McNarney, Oct. 11, 1943, both in Marshall Research File.

23. King Memo for Secretary of the Navy, etc., Nov. 4, 1943, file no. 446, General Board Records, NHD; see also Caraley, *Unification,* pp. 91–107.

24. Rowcliff to Chairman General Board, Nov. 20, 1943 General Board Records, NHD; Tompkins Memo for Marshall, Apr. 12, 1944, Marshall Research File; Hart to Secretary of the Navy, Dec. 14, 1943, file no. 446, General Board Records, NHD.

25. Marshall Memo for King, Apr. 17, 1944; King Memo for Marshall, Apr. 12, 1944, both in Marshall Research File.

26. Palmer to Wadsworth, Dec. 30, 1943, May 24, 1944, Palmer Papers.

27. Palmer to Budd, Apr. 18, 1944; Palmer to Wadsworth, May 24, 1944; "Suggested Agenda for the Committee on Post-War Policy," Apr. 15, 1944, all in Palmer Papers.

28. Marshall Memo for the Secretary of War, Apr. 17, 1944, Nelson File, War Department Records, Record Group 165, National Archives; Pogue, *Marshall, 1943–1945,* p. 364; Stimson Diary, Apr. 17 and 18, 1944, Stimson Diaries (microfilm), Yale University Library.

29. Knox to Sabath, Mar. 23, 1944; Forrestal's Office File, Secretary of the Navy Records, Record Group 80, National Archives (hereafter, Forrestal File, NA); Dillon Memo for Forrestal, May 8, 1944, Forrestal File, NHD; Warner to Knox, Dec. 14, 1943; Forrestal to Knox, Jan. 20, 1940; Knox to Cox, Mar. 9, 1944, all in Frank Knox Papers, Library of Congress; Knox to Vinson, Mar. 11, 1944, Forrestal File, NA.

30. Duffield to Forrestal, Apr. 26, 1944, Forrestal File, NA; Eberstadt to Forrestal, Apr. 7, 19, and 29, 1944, Eberstadt Papers.

31. U.S. Congress, House, *Proposal to Establish a Single Department of Armed Forces, Hearings,* 78th Cong., 2d sess., 1944, pp. 14–15, 31, 43, 91, 111; Davison to Nelson, Apr. 27, 1944; Remarks by General Nelson, Apr. 21, 1944; Patterson Statement, draft, Apr. 25, 1944, all in Nelson File, War Department Records; U.S. Congress, House.

32. Excerpts of Editorial Comments on Single Department Hearings, May 15, 1944, Robert P. Patterson Papers, Library of Congress (hereafter, Patterson Papers); *New York Times,* Apr. 27, 1944, p. 22.

33. Eberstadt to Taylor, May 9, 1944, Eberstadt Papers; Caraley, *Unification,* p. 28ff.

34. U.S. Congress, House, *Single Department Hearings,* 78th Cong., 2d sess., 1944, pp. 122–30.

35. Eberstadt to Shields, Apr. 28, 1944, Eberstadt Papers; Smith Memo for President, May 2, 1944, Smith Papers, FDRL; Miles to Hoelscher, Sept. 25, 1944; Hoelscher to Smith, June 17, 1944, series 39.32, both in Bureau of the Budget Records in Record Group 51, NA.

36. Albion and Connery, *Forrestal and the Navy,* p. 128ff; *Forrestal Diaries,* p. 5–9ff.

37. *Forrestal Diaries,* pp. 3–4.

38. Truman, "Our Armed Forces Must Be Unified," *Collier's,* Aug. 26, 1944, pp. 63–64; *Forrestal Diaries,* pp. 88–89.

39. Truman, "Our Armed Forces," pp. 16, 64; *New York Times,* Aug. 18, 1944, p. 28. See also Pemberton, *Bureaucratic Politics.*

40. Forrestal to Vinson, Aug. 30, 1944, Forrestal Papers; Forrestal to Nimitz, Sept. 6, 1944, Forrestal File, NHD; *Forrestal Diaries,* p. 9; Forrestal to King, Sept. 1, 1944; Forrestal to Hepburn, Sept. 19, 1944, both in Forrestal Papers; Brown to Roper, Sept. 2, 1944; White to Chief of Naval Personnel, Sept. 6, 1944, both in Forrestal File, NA.

41. Forrestal-Herter Telephone Conversation, transcript, Apr. 9, 1945, Forrestal Papers; Forrestal to King, Feb. 8, 1945; Forrestal to Edwards, Mar. 28, 1945, both in Forrestal File, NA.

42. Forrestal Memo for President, Apr. 9, 1945; Forrestal Memo for President Truman, Apr. 20, 1945, both in Forrestal Papers; *Forrestal Diaries,* pp. 5, 46–47.

43. Truman, *Public Papers, 1945,* pp. 275–77; Nelson quoted in Catton, *War Lords,* p. 108.

44. *Forrestal Diaries,* pp. 9, 19; Schwarz, *Speculator,* p. 449.

45. *Forrestal Diaries,* p. 117.

46. Ibid., p. 73; Gates to Forrestal, Jan. 3, 1945, Forrestal File, NHD.

47. McCain to Forrestal, Apr. 28, 1945, Forrestal File, NHD; Ramsay to Sec-

retary of the Navy, Apr. 10, 1945; Russell Memo for King, July 11, 1945; Russell Analysis of JCS 749/12 (Richardson Report); Leahy Memo for Secretary of the Navy, Sept. 1945, all in Forrestal File, NA; Kalbfus to Secretary of the Navy, June 15, 1945, file No. 446, General Board Records, NHD; Forrestal Diary, entries May 10 and 27, 1945, Forrestal Diaries, Seeley Mudd Manuscript Library (not in published version of diaries).

48. Minutes of Top Policy Group Meeting No. 20, May 7, 1945, Secretary of the Navy Records, Record Group 80, National Archives.

49. Forrestal to Stimson, May 7, 1945, Forrestal File, NHD; Correa to Forrestal, Apr. 21, 1945, Forrestal File, NA; Hensel Memo for Eberstadt, June 7, 1945, Eberstadt Papers; *Forrestal Diaries,* pp. 60–61.

50. Stimson to Forrestal, May 26, 1945, copy in Eberstadt Papers; *Forrestal Diaries,* p. 60; Davis, *Postwar Defense Policy,* p. 209.

51. U.S. Congress, House, *Single Department Hearings,* 78th Cong., 2d sess., 1944, p. 124.

52. E. E. Morison to Christman, July 2, 1970, Christman Papers; Eberstadt to Taylor, May 9, 1944, Eberstadt Papers; Forrestal to Lippmann, May 6, 1944, Forrestal File, NA; *Forrestal Diaries,* pp. 7–8.

53. *Forrestal Diaries,* p. 48ff.

54. Acheson, *Creation,* p. 184; also Rogow, *James Forrestal.*

55. Forrestal's Appointment Calendar and Telephone Log for May 1, 3, 8, 10, 23, 24, and 31, 1945 for telephone conversations and personal visits with Eberstadt, Forrestal Papers.

CHAPTER 6

1. Eberstadt to Forrestal, Apr. 18, 1945, Eberstadt Papers.

2. Forrestal to Eberstadt, May 12, 1945, Eberstadt Papers; Forrestal's Appointment Calendar and Telephone Logs for April 16, 17, and 19, May 1, 3, 8, 10, 23, 24, and 31, 1945, Forrestal Papers.

3. Eberstadt to Forrestal, Apr. 18, May 17, 1945, Eberstadt Papers.

4. *Forrestal Diaries,* p. 57; Morison to Christman, July 2, 1970, Christman Papers.

5. Eberstadt Memo of Meeting, June 8, 1945; Eberstadt Memo for Forrestal, July 27, 1945; Eberstadt to Dennison, Sept. 19, 1949, all in Eberstadt Papers.

6. Dillon to Christman, June 12, 1945, Christman Papers; Ferdinand Eberstadt interview with Christman, July 18, 1969, Princeton.

7. *Forrestal Diaries,* pp. 57, 62; Willett Memo for Topical File, June 23, 1945, Eberstadt Papers.

8. Eberstadt's Investigation Record, June 7, 1945; Memo of Meeting be-

tween Eberstadt and Willett, Fox, and Russell, June 8, 1945, both in Eberstadt Papers.

9. Forrestal to Stimson, June 18, 1945, Forrestal File, NHD; Eberstadt Investigation Record, June 7, 1945; Memo of Meeting, June 8, 1945; Forrestal to Eberstadt, June 19, 1945, all in Eberstadt Papers; *Forrestal Diaries,* p. 63.

10. Morison to Christman, July 2, 1970, Christman Papers; Eberstadt Memo of Meeting, June 8, 1945, Eberstadt Papers; U.S. Congress, Senate, Committee on Naval Affairs, *Report . . . on Unification of the War and Navy Departments,* 79th Cong., 1st sess., 1945, p. 16.

11. Morison to Christman, July 2, 1970, Christman Papers; Memo for Mr. Willett's Personal File, June 13, 1945, Eberstadt Papers.

12. Morison to Christman, July 2, 1970, Christman Papers; Eberstadt to King, June 22, 1945; Eberstadt Memo for Forrestal, June 23, 1945, both in Eberstadt Papers; Forrestal to King, Aug. 18, 1945, Forrestal File, NA.

13. Willett Memo for Topical File, June 23, 1945; Willett Memos for Eberstadt, July 2, and June 27, 1945; Memo for Mr. Willett's Personal File, June 13, 1945; Eberstadt Memo for Forrestal, June 23, 1945, all in Eberstadt Papers.

14. Forrestal to Bureaus, Offices and Shore Establishments, June 26, 1945; Willett Memo for Personnel File, June 15, 1946, both in Eberstadt Papers. Complete list of the Eberstadt Study Group personnel is in Eberstadt to Taylor, Oct. 5, 1945, Forrestal File, NA.

15. Catron Memo for Willett, July 18, 1945, Eberstadt Papers; Milton Katz letter to the author, Apr. 16, 1984.

16. See Waldo, *Administrative State,* pp. 145–46; Karl, *Charles E. Merriam,* p. 288; Elliott, *Need for Constitutional Reform,* p. 219n1; Eberstadt borrowed Herring's works from the Library of Congress (Gooch to Willett, Aug. 1, 1945, Eberstadt Papers).

17. Lyons, Morton, and Masland, *Schools for Strategy,* p. 37; Hyneman, *Study of Politics,* pp. 120–23; Herring, *Impact of War,* p. 214; Herring, *Public Administration,* pp. viii, 378.

18. Herring, *Impact of War,* pp. 192–206, 202 (quote).

19. Ibid., pp. 171–81, 226–31; Herring, *Public Administration,* pp. 314–15, 380; see also Elliott, *Need for Constitutional Reform.*

20. Herring, *Impact of War,* pp. 159, 238–41, and *Public Administration,* pp. 362–75, 383–85.

21. Herring Memo for Eberstadt, Aug. 9, 1945; Herring to Willett, July 18, 1945; Willett Memo for Eberstadt, July 27, 1945; Eberstadt's Investigation Record for July–Aug., 1945, all in Eberstadt Papers; Herring, *Impact of War,* p. 139.

22. Eberstadt Memo to self, Aug. 24, 1945; Eberstadt interview with Paul Nitze, July 18, 1945; Eberstadt Investigation Record for June 12–Aug. 11, 1945; Eberstadt Memo of Thoughts, July 12, 1945, all in Eberstadt Papers.

23. Eberstadt to Baruch, Nov. 19, 1943, Baruch Papers.

24. Morison to Christman, July 2, 1970; Gilmore to Christman, Sept. 14, 1970; Morse to Christman, Nov. 2, 1970, all in Christman Papers; Connery to Eberstadt, Nov. 28, 1945, Eberstadt Papers; Katz letter to author, Apr. 16, 1984.

25. Eberstadt's Investigation Record, July–Aug., 1945, Eberstadt Papers; Forrestal to Stimson, July 25, 1945, Forrestal File, NHD; Patterson to Eberstadt, July 23, 1945; Eberstadt Memo to Forrestal, Sept. 3, 1945; Eberstadt to Forrestal, July 25, 1945, all in Eberstadt Papers; Patterson to Forrestal, Sept. 4, 1945, Forrestal File, NHD.

26. Eberstadt's Notes, Aug. 1, 1945; Eberstadt's Notes for Discussion with Mr. Woodrum, July 26, 1945; Eberstadt to Forrestal, Sept. 25, 1945, all in Eberstadt Papers.

27. Eberstadt's Notes, July 20, 1945; Eberstadt to Forrestal, Sept. 25, 1945, both in Eberstadt Papers.

28. Eberstadt to Forrestal, Aug. 6, 1945; Lehman to Eberstadt, June 23, 1945; President Truman to Eberstadt, June 29, 1945; Eberstadt to Truman, July 6, 1945; Vardaman to Forrestal, Aug. 14, 1945; Eberstadt Memo of Thoughts on Lend-Lease, July 12, 1945, all in Eberstadt Papers; Forrestal to Truman, Aug. 9, 1945, Forrestal Papers.

29. Eberstadt Memo to Forrestal, Oct. 5, 1945; Eberstadt to Dennison, Sept. 19, 1949, both in Eberstadt Papers; see also Burch, Jr., *Elites in American History,* 3:18ff, 110ff.

30. Searls to Eberstadt, Aug. 23, 1945, Eberstadt Papers.

31. Hancock to Eberstadt, Aug. 24, 1945; Williams to Eberstadt, Sept. 10, 1945; Notes on interview with John Rockefeller, Aug. 7, 1945; Eberstadt Notes on Vannevar Bush interview, July 26, 1945; Memo of Talk with Novick, July 12, 1945, all in Eberstadt Papers; see also Schwarz, *Speculator,* pp. 253, 263, 442, 544–45; and Hammond, *Organizing for Defense,* for interpretation that the NSC idea came from British organization.

32. Root, Jr., to Eberstadt, Sept. 20, 1945; Emmerich to Eberstadt, Sept. 7, 1945; Groves to Eberstadt, Sept. 6, 1945; Williams to Eberstadt, Sept. 10, 1945; Eberstadt Notes on Bush interview, July 26, 1945; Memo of Conference with Conant, Aug. 25, 1945; Bard to Eberstadt, Aug. 25, 1945, all in Eberstadt Papers. (Names from the Good Man List dominated PACH, including Bard, Charles Dawes, Marshall Field, and John Lord O'Brian.)

33. Baldwin to Eberstadt, Aug. 17, 1945; Wilson to Eberstadt, Sept. 14, 1945; Eberstadt Memo on Sarnoff interview, Aug. 31, 1945, all in Eberstadt Papers.

34. U.S. Congress, Senate, *Report on Unification,* 79th Cong., 1st sess., 1945, p. 31.

35. Ibid., pp. 20, 42.

36. Ibid., pp. 11–12.

37. Ibid., pp. 2, 36–37, 50.

38. McCormick, "Drift or Mastery?" p. 326; U.S. Congress, Senate, *Report on Unification,* 79th Cong., 1st sess., 1945, p. 17.

39. Eberstadt Memo of Thoughts in Connection with the Report, June 28, 1945; Eberstadt to Forrestal, Oct. 5, 1945, Eberstadt Papers; Chambers, *To Raise an Army,* p. 204.

40. Eberstadt Memo to Drafting Committee, Aug. 27, 1945; Eberstadt to Forrestal, Sept. 25, 1945, both in Eberstadt Papers; Baruch quoted in Schwarz, *Speculator,* p. 545.

41. Eberstadt to Forrestal, Sept. 25, 1945, Eberstadt Papers; U.S. Congress, Senate, *Report on Unification,* 79th Cong., 1st sess., 1945, pp. 8–9, 11.

42. Eberstadt's Notes, July 20, 1945; Eberstadt Memo of Thoughts, July 12, 1945; Eberstadt to Forrestal, Sept. 25, 1945, both in Eberstadt Papers; U.S. Congress, Senate, *Report on Unification,* 79th Cong., 1st sess., 1945, pp. 33, 37.

CHAPTER 7

1. MacGregor to Christman, Nov. 24, 1970, Christman Papers.

2. Smith Memo for President, May 22, 1946, President's Secretary's File, Truman Papers; Smith Diary, May 22, 1946, Smith Papers, FDRL.

3. Forrestal to Krock, Oct. 17, 1945, Krock Papers; Forrestal to Watson, Nov. 8, 1945, Forrestal Papers; Eberstadt to Forrestal, Dec. 20, 1945, Eberstadt Papers.

4. *Forrestal Diaries,* p. 145; Forrestal Memo to Eberstadt, Sept. 24, 1945, Forrestal Papers.

5. Forrestal to Eberstadt, Oct. 6, 1945, Eberstadt Papers.

6. Eberstadt Brief of Remarks Off the Record to Luncheon, Nov. 2, 1945, typescript, Eberstadt Papers.

7. Forrestal to Sulzberger, Sept. 28, 1945; Forrestal to Rosenman, Oct. 13, 1945, both in Forrestal Papers; Strauss to Forrestal, Sept. 14, 1945, Forrestal File, NA; Eberstadt to Forrestal, Sept. 27, Oct. 5, 1945, Eberstadt Papers.

8. *Forrestal Diaries,* p. 50; Rogow, *James Forrestal,* pp. 330–31; see also Haynes, *Awesome Power;* Yergin, *Shattered Peace.*

9. Blum, ed., *Wallace Diary,* pp. 487, 491; Smith Conference with the President, Oct. 30, 1945, Smith Papers, HSTL; see also Haynes, *Awesome Power;* Pemberton, *Bureaucratic Politics.*

10. Elsey's Notes on Reorganization, Oct. 15, 1945, George M. Elsey Papers, HSTL; Pemberton, *Bureaucratic Politics,* p. 281; Truman, *Memoirs,* 2:46; Smith Conference with the President, Oct. 30, 1945, Smith Papers, HSTL.

11. Elsey Memo, Nov. 2, 1945; "Roosevelt's Views on the Unification of the Armed Forces," Memo, Dec. 6, 1945, both in Elsey Papers.

12. Truman, *Memoirs,* 2:1–48; Leahy Memo, Sept., 1945; King Memo for JCS, Sept. 21, 1945; JCS Memo for the President, Oct. 16, 1945, all in Forrestal File, NA.

13. Truman to Wiley, Nov. 1, 1945, President's Official File 335A, Truman Papers; Conference with the President, Oct. 30, 1945, Smith Papers, HSTL.

14. Handy Memo to Marshall, Sept. 29, 1945, Marshall Research File; Patterson Memo for Persons, Oct. 16, 1945, Patterson Safe File, Secretary of War Records, Record Group 107, National Archives; Collins to Christman, Oct. 22, 1970, Christman Papers; Chandler and Galambos, eds., *Papers of Dwight David Eisenhower,* 6:119ff (hereafter, *Eisenhower Papers*); U.S. Congress, Senate, *Department of Armed Forces Hearings on S. 84 and S. 1482,* 79th Cong., 1st. sess., 1945, pp. 10–24, 49ff.

15. Willett to Eberstadt, Nov. 16, 1945, Eberstadt Papers; U.S. Congress, Senate, *Department of Armed Forces Hearings,* 79th Cong., 1st sess., 1945, p. 360.

16. Willett Memo for Eberstadt, Nov. 9, 1945, Eberstadt Papers; U.S. Congress, Senate, *Department of Armed Forces Hearings,* 1945, 79th Cong., 1st sess., pp. 233, 308, 344.

17. Redding to Truman, Nov. 9, 1945; Hassett to Redding, Nov. 9, Dec. 18, 1945, President's Official File 335A, Truman Papers; see also Mrozek, "Peace through Strength" (Ph.D. diss.), ch. 2.

18. Eliot to Forrestal, Nov. 19, 1945, Forrestal File, NA; Rabb Memo to Eberstadt, Nov. 14, 1945, Eberstadt Papers; AAF Special Organization Plans, "Second Weekly Letter on Single Department of Armed Forces," Nov. 16, 1945, Carl Spaatz Papers, Library of Congress.

19. Forrestal to Eberstadt, Oct. 6, 1945; Eberstadt Memo for Forrestal, Oct. 5, 1945, both in Eberstadt Papers.

20. Rabb Memo to Eberstadt, Oct. 23, 1945; Willett Memo to Eberstadt, Oct. 23, 1945; Rabb to Eberstadt, Oct. 31, 1945, all in Eberstadt Papers.

21. Swope to Eberstadt, Oct. 29, 1945; Eberstadt to Swope, Nov. 15, 1945; Eberstadt to Vinson, Oct. 20, 1945, all in Eberstadt Papers.

22. Gates to Forrestal, Oct. 25, 1945; Memo of Meeting in Secretary of the Navy's Office, Oct. 18, 1945; Forrestal Memo to Gates, Hensel, Eberstadt, etc., Oct. 20, 1945, all in Forrestal File, NA; see also Lippmann to Gates, Nov. 7, 1945; Eberstadt's corrections on Hensel's draft, Nov. 20, Dec. 20, 1945, both in Eberstadt Papers.

23. U.S. Congress, Senate, *Department of Armed Forces Hearings,* 79th Cong., 1st sess., 1945, pp. 241–68; Hensel to Eberstadt, Nov. 7, 1945; Rabb to Eberstadt, Nov. 8, 1945, both in Eberstadt Papers.

24. Harrod to Eberstadt, Nov. 5, 1945; Willett Memo to Eberstadt, Nov. 30, 1945; Eberstadt to Catron, Nov. 13, 1945; Catron to Eberstadt, Dec. 8, 1945; *The Navy Plan for National Security,* brochure, Dec. 12, 1945, all in Eberstadt Papers; Harrod to Christman, Jan. 15, 1971, Christman Papers. The navy brochure reached

the White House on Dec. 11, 1945 (Hensel to Rosenman, Dec. 11, 1945, Samuel I. Rosenman Papers, HSTL).

25. Eberstadt to Forrestal, Sept. 27, Oct. 5, 22, and 29, Nov. 15 and 16, 1945; Eberstadt to Hensel, Dec. 5, 1945, all in Eberstadt Papers; Forrestal to Eberstadt, Nov. 5, 1945; Eberstadt to Duffield, Oct. 19, 1945; Eberstadt to Overfelt, Nov. 5, 1945; Overfelt to Eberstadt, Nov. 9, 1945, all in Forrestal File, NA.

26. Eberstadt to Forrestal, Nov. 15 and 18, 1945, Eberstadt Papers. The SCOROR concept developed during a meeting between Hensel, Radford, and Vinson and reflected their views, but Eberstadt synthesized these and presented the idea to Forrestal.

27. Eberstadt to Forrestal, Nov. 18, 1945; Eberstadt Speech, draft, Nov. 1945, both in Eberstadt Papers.

28. Elsey Notes, Nov. 1, 1945; copy of Miller Memo to All Flag Officers, Nov. 14, 1945, both in Elsey Papers; Ayers Diary, entry Dec. 19, 1945, Eben A. Ayers Papers, HSTL; Smith Conference with President, Oct. 30, 1945, Smith Papers, HSTL.

29. Baetich to the President, Dec. 2, 1945; HST penciled note on Sheldon Clark to President, Dec. 1, 1945, President's Official File 335A, both in Truman Papers, Forrestal to Hillman, Nov. 28, 1945; Forrestal to Green, Nov. 29, 1945; Forrestal-Lewis Telephone Conversation, transcript Dec. 3, 1945; Forrestal-Dillon Telephone Conversation, transcript, Nov. 20, 1945, all in Forrestal Papers; Green to Truman, Nov. 29, 1945, Forrestal File, NA; Sheehy to Connelly, Dec. 3, 1945; Truman Papers.

30. Smith Conference with the President, Dec. 13, 1945, Smith Papers, HSTL; Eberstadt to Millis, Nov. 13, 1945; Eberstadt to Pearson, Nov. 16, 1945; Willson to Eberstadt, Nov. 21, 1945; Eberstadt to Marshall, Oct. 30, 1945, all in Eberstadt Papers; Transcript of Swing radio broadcast, Dec. 14, 1945, Rosenman Papers, HSTL; *New York Times,* Nov. 5 and 20, 1945; *New York Herald-Tribune,* Dec. 13, 1945; Smith Memo for President, Nov. 30, 1945, Smith Papers, HSTL.

31. Smith Memo for the President, Dec. 6, 1945; Smith Conference with the President, Dec. 13, 1945, both in Smith Papers, HSTL; Rosenman Memo to the President, Nov. 19, 1945, Rosenman Papers, HSTL.

32. Elsey Notes, Nov. 1, Dec. 28, 1945; Elsey Summary of Eberstadt Report, Nov. 8, 1945, both in Elsey Papers; Rosenman, quoted in Haynes, *Awesome Power,* p. 298n19; *Forrestal Diaries,* p. 89; Smith Conference with President, Dec. 13, 1945, Smith Papers, HSTL.

33. Elsey Oral History, p. 14ff, HSTL; *Lilienthal Journals,* 2:209.

34. Elsey Notes on Reorganization, Oct. 13 and 15, Dec. 17, 18, and 19, 1945; Elsey Memo for Clifford, Nov. 29, 1945; Status of Eberstadt Report, Oct. 15, 1945, all in Elsey Papers.

35. Elsey draft, Dec. 13, 1945, Elsey Papers. Clifford Memo for Judge Rosenman, Dec. 13, 1945, Rosenman Papers, HSTL.

36. Eberstadt to Hensel, Dec. 28, 1945, Eberstadt Papers; Elsey Notes on Reorganization, Dec. 14, 1945; Rosenman Memo for Clifford, Dec. 17, 1945, Elsey Papers; see also Caraley, *Unification.*

37. Miles Memo for Judge Rosenman on Proposed Insert in Army-Navy Unification Message, Dec. 15, 1945, Rosenman Papers; James L. Sundquist Oral History, 1964, pp. 3–35, HSTL.

38. Truman, *Public Papers,* 1945, pp. 546–60; Caraley, *Unification,* p. 83.

39. Truman, *Public Papers,* 1945, pp. 547, 551–52, 558–59.

40. Ibid., pp. 553–54, 560.

41. U.S. Congress, Senate, *Report on Unification,* 79th Cong., 1st sess., 1945, pp. 2–31; Truman, *Public Papers,* 1945, pp. 553–54, 559.

42. *Forrestal Diaries,* p. 92; Truman, *Public Papers,* 1945, pp. 548, 550.

43. U.S. Congress, Senate, *Report on Unification,* 79th Cong., 1st sess., 1945, p. 16ff.

44. Truman, *Public Papers,* 1945, p. 555.

45. Ibid.; U.S. Congress, Senate, *Report on Unification,* 79th Cong., 1st sess., 1945, p. 36.

46. Petersen Memo for Rosenman, Dec. 18, 1945, Rosenman Papers, HSTL; *Forrestal Diaries,* pp. 118–20; Elsey Notes on Reorganization, Dec. 19, 1945, Elsey Papers; Leahy Diary, entry Feb. 8, 1946, William D. Leahy Diaries, Library of Congress.

47. Eberstadt to Forrestal, Dec. 20, 1945, Eberstadt Papers; Forrestal-Eberstadt Telephone Conversation, Dec. 19, 1945, transcript, Forrestal File, NA.

48. Eberstadt to Forrestal, Dec. 20, 1945, Eberstadt Papers; U.S. Congress, Senate, *Report on Unification,* 79th Cong., 1st sess., 1945, pp. 25–26; Forrestal-Eberstadt Telephone Conversation, Dec. 19, 1945, transcript, Forrestal File, NA.

49. Forrestal-Eberstadt Telephone Conversation, Dec. 19, 1945, transcript, Forrestal File, NA; Eberstadt to Forrestal, Dec. 20, 1945, Eberstadt Papers.

50. Forrestal-Bard Telephone Conversation, Dec. 20, 1945, transcript, Forrestal Papers.

51. Ayers Diary, entry Dec. 19, 1945, Ayers Papers; *Forrestal Diaries,* pp. 88–89; Sullivan-Taylor Telephone Conversation, Nov. 21, 1945, transcript, Forrestal File, NA.

52. Truman to Forrestal, Dec. 17, 1945, Forrestal File, NA.

53. Forrestal to Ridder, Dec. 21, 1945; Forrestal Memo for Duffield, Dec. 23, 1945; Forrestal-Bard Telephone Conversation, Dec. 20, 1945, transcript, all in Forrestal Papers.

54. Eberstadt-Forrestal Telephone Conversation, Jan. 11, 1946, transcript, Forrestal File, NA; Forrestal to Cahill, Jan. 5, 1946, Forrestal Papers.

55. *Forrestal Diaries,* p. 127.

56. Ibid., pp. 110, 127; Eberstadt-Forrestal Telephone Conversation, Mar. 7, 1946, transcript, Forrestal File, NA.

57. Forrestal-Lewis Douglas Telephone Conversation, Jan. 10, 1946, transcript, Eberstadt Papers; *Forrestal Diaries*, pp. 105–106.

CHAPTER 8

1. Eberstadt to Forrestal, Jan. 19, 1946, Eberstadt Papers.

2. *Truman Memoirs*, 1:553; *Forrestal Diaries*, p. 130; Forrestal to Admiral Robinson, Jan. 7, 1946, Forrestal Papers. Strauss, *Men and Decisions*, p. 157, suggested that Forrestal had presidential aspirations.

3. Forrestal to Patterson, Dec. 12, 1945; Forrestal-Lewis Douglas Telephone Conversation, transcript, Jan. 10, 1946, Eberstadt Papers; Forrestal to Krock, Oct. 17, 1945, Krock Papers; *Eisenhower Papers*, 7:701ff.

4. SCOROR Admirals were Arthur W. Radford, Thomas H. Robbins, H. B. Miller, and Tomlinson. Eberstadt Study Group members included Catron, Willett, Rabb, Carr, and Richard Moser.

"Organization of Personnel Working on Merger Question," Jan. 5, 1946; Forrestal-McCormack Telephone Conversation, transcript, Feb. 19, 1946, Forrestal File, NA. Catron to Eberstadt, Jan. 5 and 11, Feb. 21 and 26, 1946; Radford to Eberstadt, Mar. 6, 1946; and "Status of Army-Navy Merger Problem as Reported by the Commandants of the Naval Districts," Feb. 21, 1946, all in Eberstadt Papers.

5. Catron Memo for Secretary of the Navy, Jan. 16, 1946, Forrestal File, NA; Catron to Eberstadt, Jan. 11, 1946; Bowen to Eberstadt, Feb. 4, 1946, both in Eberstadt Papers.

6. Senator Austin to Eberstadt, Jan. 11, 1946; Thomas-Forrestal Telephone Conversation, transcript, Jan. 11, 1946; Rabb Memo for Eberstadt, Nov. 7, 1945; Catron to Eberstadt, Jan. 11, 1946, all in Eberstadt Papers.

7. Eberstadt to Senator Austin, Jan. 21, 1946; Austin to Eberstadt, Jan. 10, 1946; Rabb Memo for Eberstadt, Nov. 19, 1945; Forrestal-Douglas Telephone Conversation, transcript, Jan. 10, 1946; Eberstadt to Forrestal, Jan. 11, 1946, all in Eberstadt Papers; Hidalgo Memo for Secretary of the Navy, Jan. 12, 1946; Forrestal to Hensel, Jan. 12, 1946, both in Forrestal File, NA.

8. *Lilienthal Journals*, 2:28; Blum, ed., *Wallace Diary*, p. 562.

9. *Forrestal Diaries*, pp. 134, 143–46.

10. ALUSNA (Naval Attaché), Mexico City to Forrestal, Feb. 19, 1946, Forrestal Papers; Radford to Eberstadt, Feb. 19, 1946; Catron Memos to Eberstadt, Feb. 21 and 26, 1946, all in Eberstadt Papers.

11. Eberstadt-Forrestal Telephone Conversation, transcript, Mar. 7, 1946, Forrestal File, NA; Eberstadt to Forrestal, Mar. 5, 1946; Eberstadt to Catron, Mar. 5, 1946; Eberstadt to Austin, Jan. 21, 1946, Eberstadt Papers. For the FBI investigation of Senator Thomas, see Kenneth O'Reilly, "A New Deal for the FBI," *Journal of American History* 69 (Dec., 1982): 650.

12. Eberstadt-Forrestal Telephone Conversation, transcript, Mar. 7, 1946, Forrestal File, NA; Forrestal Memos to President, Mar. 8 and 12, 1946, Forrestal Papers; *Forrestal Diaries,* p. 149; Forrestal testimony, in U.S. Congress, Senate, Committee on Naval Affairs, *Unification Hearings on S. 2044*, 79th Cong., 2d sess., 1946, pp. 38–39, 43; Truman to Smith, Apr. 10, 1946, President's Secretary's File 145, Truman Papers.

13. Forrestal to Admiral Bratton, Mar. 11, 1946, Forrestal File, NA; *Forrestal Diaries,* p. 149.

14. Forrestal Memo for the President, Apr. 20, 1945, Forrestal Papers; *Forrestal Diaries,* pp. 46–47.

15. Forrestal to Baldwin, Mar. 14, 1946, Eberstadt Papers; Forrestal's ALNAV, Mar. 28, 1946, Forrestal Papers; *Forrestal Diaries,* p. 155.

16. Leach Memo for Spaatz, Dec. 13, 1945; Knerr to Spaatz, Feb. 26, 1946, Spaatz Papers.

17. Spaatz to Key, Feb. 8, 1946; Spaatz Remarks Made before the Aviation Writers Association Dinner, Mar. 14, 1946, both in Spaatz Papers.

18. Spaatz to Warner, Feb. 7, 1946; Bagby (Spaatz's aide) to Beattie, Mar. 19, 1946; Cooper to Spaatz, Jan. 2, 1946, all in Spaatz Papers; Spaatz, "Air Power in the Atomic Age," *Collier's* 5 (Dec. 8, 1945): 11–12; "The Sky's the Limit," Radio Script, AAF Public Information Division, Secretary of the Air Force Records, Record Group 340, National Archives; on Forrestal-monitored air force activity, see Secret Report from Commandant, 12th Naval District, to Forrestal, Jan. 1, 1946, Forrestal File, NA.

19. Harriman quoted in Lovett to Spaatz, Mar. 27, 1946, Spaatz Papers; Truman to Crain, Dec. 17, 1945, President's Official File 335A, Truman Papers.

20. Symington to Forrestal, Mar. 15, 1946; "Mobilizing Grass Roots Support for a Single Department of National Defense and an Adequate Air Force," Nov. 5, 1945; Remarks made by General Spaatz, Mar. 14, 1946; Spaatz to Lovett, Apr. 4, 1946, Spaatz Papers; Legere to author, Nov. 9, 1984; Jurika, Jr., ed., *From Pearl Harbor to Vietnam: Memoirs of Admiral Arthur W. Radford* (hereafter, *Radford Memoirs*), p. 82; Huie, *Case against the Admirals;* William Bradford Huie, interview with author, Mar. 11, 1984; Eric Sevareid Newscast, typescript, Mar. 31, 1946, Spaatz Papers.

21. Radford Memo on Possible Concessions, Mar. 11, 1946; Hidalgo Memo to Secretary of the Navy, Mar. 8, 1946; Patterson to Forrestal, Mar. 11, 1946, Eberstadt Papers; Hidalgo Memo for Patterson, Mar. 9, 1946, Forrestal File, NA.

22. Forrestal to the President, Mar. 12, 1946; Eberstadt to Forrestal, Aug. 12, 1946, both in Forrestal Papers; Eberstadt Memo Re Patterson-Forrestal Negotiations, May 20 and 29, 1946; Eberstadt to Eliot, Apr. 17, 1946; Eberstadt to Forrestal, Apr. 1, 1946, all in Eberstadt Papers.

23. Eberstadt to Senator Bridges, Apr. 16 and 26, 1946; Senator Hart to Eberstadt, May 14, 1946; Eberstadt to Duffield, Apr. 17, 1946; Eberstadt Draft

of Proposed Minority Report, Apr. 16, 1946; Eberstadt to Radford, May 1, 1946; Eberstadt to Nash, May 1, 1946; Eberstadt to Forrestal, May 2, 1946; Walsh to Eberstadt, Apr. 8, 1946, all in Eberstadt Papers; Nunn Memo to Judge Advocate General, Apr. 25, 1946, Forrestal File, NA.

24. Advice to Adler, in Eberstadt to Krock, Oct. 17, 1946; McCloy to Eberstadt, telegram, Apr. 1, 1946; Eberstadt to Forrestal, Oct. 29, 1946; Eberstadt to McCloy, Sept. 27, 1946; Eberstadt to Forrestal, telegram, Apr. 1, 1946; Memorandum of Eberstadt, Lovett, McCloy Meeting, Sept. 26, 1946; Gates to McCloy, Dec. 6, 1946; Lovett to Eberstadt, Nov. 18, 1946, all in Eberstadt Papers; McCloy to Patterson, Oct. 16, 1946, Patterson Safe File, Secretary of War Records, Record Group 107, National Archives.

25. John A. Kennedy Oral History, 1978; Robert L. Dennison Oral History, 1972; John Abbott Oral History, 1970, pp. 102–107, all in HSTL.

26. William R. Smedberg Oral History, 1979, 1:314, U.S. Naval Institute, Annapolis, Md.

27. Nash Memos to Eberstadt, Apr. 29, May 31, June 5, and 26, 1946, Eberstadt Papers.

28. Eberstadt Draft of Testimony for Senate Naval Affairs Committee, May 9, 1946, Eberstadt Papers.

29. Symington Memos for Secretary of War Patterson, Apr. 22, Aug. 29, Nov. 20, 1946, Patterson Safe File, Secretary of War Records, Record Group 107, National Archives; Eberstadt to Symington, Oct. 14, 1946, Eberstadt Papers; Wolk, *Planning and Organizing the Postwar Air Force,* p. 93.

30. Chapman to the President, June 18, 1946, President's Official File 335A; Smith Memo to President, May 22, 1946, President's Secretary's File 145, both in Truman Papers; Minutes of Conference with President, May 22, 1946, Smith Papers, HSTL.

31. See *Forrestal Diaries,* p. 62, for definition of militarized civilian; also Raymond, *Power at the Pentagon,* p. 321; Barnet, *Roots of War;* Lane, *Armed Progressive.*

32. Smith Memo for the President, May 22, 1946, President's Secretary's File 145, Truman Papers; Kenney to Eberstadt, Sept. 27, 1946, Eberstadt Papers; *Truman Memoirs,* 2:46–52.

33. Eberstadt to Lovett, Jan. 20, 1947; Eberstadt Memo to self, Sept. 13, 1946, both in Eberstadt Papers. For Eberstadt's role on the Baruch Atomic Energy study group, see Schwarz, *Speculator,* pp. 490–507, and Gregg Herken, *The Winning Weapon: The Atomic Bomb and the Cold War, 1945–1950* (New York: Alfred A. Knopf, 1980), pp. 160–90.

34. Sherman-Eberstadt Telephone Conversation, in Admiral Robbins Memo for the Secretary of the Navy, Jan. 24, 1947, Forrestal File, NA; Admiral Sherman to Eberstadt, May 16, 1946, Eberstadt Papers.

35. Connor Memo for the Secretary of the Navy, Nov. 25, 1946, Forrestal File, NA.

36. Truman to Forrestal, Jan. 8, 1947, Forrestal File, NA; "Hazards of Merger," in Foskett Memo for the President, Jan. 14, 1947, President's Secretary's File 145, Truman Papers; Smedberg to Robbins, Jan. 11, 1947, Forrestal File, NA; Elsey Notes, May 12, 1947, Elsey Papers; also, Keiser, *Marine Corps and Unification.*

37. Forrestal-Gurney Telephone Conversations, transcript, Mar. 6, May 12, 1947; Eberstadt-Forrestal Telephone Conversation, transcript, Mar. 31, 1947; Forrestal to Matter, Jan. 6, 1947; Forrestal to Dilworth, Jan. 17, 1947, all in Forrestal Papers; Forrestal to Gates, Jan. 21, 1947; Forrestal to Hensel, Jan. 21, 1947, both in Forrestal File, NA.

38. *Forrestal Diaries,* pp. 294–95; Forrestal to Gurney (unsent), July 1, 1947, Forrestal File, NA.

39. Title I of National Security Act of 1947 in Wolk, *Planning and Organizing the Postwar Air Force,* Appendix 9, pp. 276–80.

40. Titles II and III, in ibid., pp. 280–92.

41. Forrestal to his son Michael, Jan. 19, 1947, Forrestal Papers; Eberstadt to Lovett, Jan. 20, 1947, Eberstadt Papers.

42. Truman to Daniels, Jan. 21, 1947, President's Official File 335A, Truman Papers; Donovan, *Conflict and Crisis,* p. 140; Truman, *Memoirs,* 2:46–52; see also Pemberton, *Bureaucratic Politics.*

43. Donovan, *Conflict and Crisis,* p. 265; Hammond, *Organizing for Defense,* p. 233; Caraley, *Unification,* pp. 135–36.

44. *Forrestal Diaries,* pp. 294–95; Nimitz incident, in Smith Conference with the President, Feb. 28, 1946, Smith Papers, HSTL.

45. Coletta, *Navy and Unification,* p. 27; Donovan, *Conflict and Crisis,* p. 310; see also Haynes, *Awesome Power.*

46. Forrestal to Hoyt, July 29, 1947, Forrestal Papers; *Forrestal Diaries,* p. 300.

CHAPTER 9

1. Forrestal's Remarks at the Team Unification Meeting, Nov. 20, 1947, Forrestal Papers; *Forrestal Diaries,* 298–99; Forrestal to Nitze, July 21, 1947, Eberstadt Papers.

2. Rearden, *Office of Secretary of Defense,* pp. 29, 59.

3. Ibid., pp. 59–67; Isaacson and Thomas, *Wise Men,* p. 184ff. Eberstadt's Good Man List, 1949, contained the names of Leva, McNeil, William Jackson, and Kenneth Johnson (Good Man List, Eberstadt Papers).

4. Strauss to Christman, July 8, 1970, Christman Papers.

5. Forrestal to Morgan H. Hoyt, July 29, 1947, Forrestal Papers; Eberstadt to Forrestal, July 23, 1947, Eberstadt Papers; Perez and Willett, *Will to Win,* pp. 65–73.

6. Truman, *Memoirs,* 2:52.

7. Eberstadt to Forrestal, July 23, Aug. 6, and 18, 1947, Eberstadt Papers;

Eberstadt to Leva, Aug. 13, 1947; Rearden, *Office of the Secretary of Defense,* p. 95.

8. *Forrestal Diaries,* pp. 307, 313. See also Rearden, *Office of the Secretary of Defense.*

9. Truman, *Memoirs,* 2:274; *Forrestal Diaries,* pp. 303–305; Forrestal to Nimitz, Apr. 2, 1947, Forrestal Papers; Eberstadt to Forrestal, Aug. 13, 1947, Eberstadt Papers.

10. Eberstadt to Forrestal, Aug. 13, 1947, Eberstadt Papers.

11. Stevens to Forrestal, Sept. 30, 1947, copy in Eberstadt Papers. See also Gardner, *Architects of Illusion.*

12. Forrestal to Dodds, Oct. 16, 1947, Eberstadt Papers; on Hayek, see Karl, *Charles E. Merriam,* p. 297.

13. Eberstadt to Forrestal, Nov. 7, 1947, also Oct. 21, 1947, and Jan. 27, 1948, all in Eberstadt Papers.

14. Eberstadt to Forrestal, Oct. 21, Nov. 7, 1947, Jan. 27, 1948; Eberstadt to Taft, Dec. 1, 1947; Eberstadt Draft of Comments for Brewster Committee, Oct. 22, 1947; Eberstadt Speech at *Princetonian* Annual Banquet, 1948, all in Eberstadt Papers.

15. Eberstadt to Lovett, Aug. 18, 1947; Eberstadt to Forrestal, Aug. 18, Sept. 5, and Nov. 6, 1947; Paget to Leva, Aug. 12, 1947, all in Eberstadt Papers.

16. Baruch to Eberstadt, Oct. 8, 1947; Eberstadt to Forrestal, Aug. 18, Sept. 23, 1947; Eberstadt's NSRB Report File, Oct. 2, 1947, all in Eberstadt Papers.

17. Truman, *Memoirs,* 2:60; Eberstadt to Forrestal, Oct. 2, 1947; Eberstadt's NSRB Report Diary, Oct. 2, 1947, both in Eberstadt Papers. See also Cuff, "Eberstadt, NSRB," pp. 45–46.

18. *Forrestal Diaries,* pp. 329–33, passim; Cuff, "Eberstadt, NSRB," pp. 37–52; Rearden, *Office of the Secretary of Defense,* pp. 130–31; Kinnard, *Secretary of Defense,* p. 22.

19. Eberstadt to Forrestal, Nov. 7, 1947; Eberstadt to Hill, Nov. 26, 1947; Eberstadt to Taft, Dec. 1, 1947; Patterson to Eberstadt, Mar. 2, 1948, all in Eberstadt Papers. See also Isaacson and Thomas, *Wise Men,* p. 428.

20. *Lilienthal Journals,* 2:260–62; *Forrestal Diaries,* p. 335; Rearden, *Office of the Secretary of Defense,* pp. 170–273; Forrestal's Greece-Personnel List, ONI Memo on Greek king, Apr. 2, 1947, Forrestal Papers.

21. Eberstadt to Chamberlain, Feb. 24, 1948, Eberstadt Papers; see also *Forrestal Diaries,* pp. 339–65; Rearden, *Office of the Secretary of Defense,* pp. 177–208; Rogow, *James Forrestal,* pp. 26–28, 143, 281.

22. Forrestal to Eberstadt, Feb. 13, 1948, Eberstadt Papers; *Forrestal Diaries,* pp. 365–66, 372–75.

23. Forrestal to Eberstadt, Feb. 13, 1948, Eberstadt Papers; also, Pogue, *Marshall: Statesman,* pp. 150, 303.

24. Rearden, *Office of the Secretary of Defense,* pp. 36, 40, 61; Eberstadt to Forrestal, Feb. 24, 1948, Eberstadt Papers.

25. Eberstadt to Forrestal, Feb. 24, 1948, Eberstadt Papers.

26. Eberstadt to Chamberlain, Feb. 24, 1948, Eberstadt Papers; Rogow, *James Forrestal,* pp. 205, 343; *Forrestal Diaries,* pp. 382–421; Donovan, *Conflict and Crisis,* pp. 357–68.

27. Truman, quoted in Rearden, *Office of the Secretary of Defense,* pp. 280–93; *Forrestal Diaries,* pp. 374–419, passim.

28. Forrestal-Hill Telegram to Eberstadt, Mar. 19, 1948; Somervell to Eberstadt, June 17, 1944, Eberstadt Papers.

29. "Names of (Good) Men to Assist in NSRB Operations, May 7, 1948," Eberstadt NSRB Diary, Apr. 14, 18, and 22, 1948, Eberstadt Papers.

30. Krug to Hill, Apr. 7, 1948; Hill to Krug, Apr. 9, 1948; Eberstadt to Willett, Apr. 30, 1948; Eberstadt NSRB Diary, Apr. 14, 18, 22, and 23, 1948, all in Eberstadt Papers; Rearden, *Office of the Secretary of Defense,* p. 144; *Forrestal Diaries,* pp. 409–10.

31. Eberstadt NSRB Memo, May 15, 1948; Eberstadt Notes on NSRB Study, Apr. 14, 1948; Eberstadt NSRB Diary, Apr. 22–23, 1948, all in Eberstadt Papers; Donovan, *Conflict and Crisis,* pp. 140–41, 265.

32. Eberstadt Memo for Forrestal, May 27, 1948; Eberstadt NSRB Diary, May 10, 1948; "Report by F. Eberstadt to Arthur M. Hill, Chairman of the NSRB, 1948," all in Eberstadt Papers.

33. Cuff, "Eberstadt, NSRB," p. 52; Johnson to Eberstadt, June 2, 1948; Hoover to Eberstadt, Mar. 24, 1948, both in Eberstadt Papers.

34. Eberstadt to Hoover, Aug. 24, 1956, Eberstadt Papers; Moe, *Hoover Commissions,* pp. 27–28; Pemberton, *Bureaucratic Politics,* pp. 79–91; Hawley, "Herbert Hoover," in Fausold, ed., *Hoover Presidency,* pp. 101–19; Cuff, "Hoover," pp. 358–72.

35. Eberstadt to Hoover, Apr. 22, 1948, Eberstadt Papers; Moe, *Hoover Commissions,* pp. 23–30; Pemberton, *Bureaucratic Politics,* pp. 79–96; Memo Re Job for Hoover Commission, May 10, 1948, Eberstadt Papers.

36. Royall to Hoover, May 25, 1948; Symington, in Eberstadt Diary, May 19, 22, 1948; Eberstadt Memo to self, May 15, 1948; Royal to Hoover, May 25, 1948; Royal Memo to Secretary of Defense, May 21, 1948, all in Eberstadt Papers.

37. Eberstadt's Task Force on National Security Organization: Raymond B. Allen, president, University of Washington; Thomas Archer, vice-president, General Motors; Hanson W. Baldwin, military editor, *New York Times;* Chester I. Barnard, president, Rockefeller Foundation; Charles W. Cole, president, Amherst College; John Cowles, president, *Minneapolis Star and Tribune;* James Knowlson, president, Stewart-Warner Corp.; John J. McCloy, president, International Bank for Reconstruction and Development; Frederick A. Middlebush, president, University of Missouri; Robert P. Patterson, lawyer, Belknap & Webb of New York City; Lewis L. Strauss, Atomic Energy Commission; J. Carlton

Ward, Jr., chairman of the board, Fairchild Engine & Airplane Corp.; Robert E. Wood, chairman of the board, Sears, Roebuck & Company.

38. Connery to Christman, July 16, 1970, Christman Papers; Eberstadt to Krock, Aug. 27, 1948, Eberstadt Papers; *Hoover Commission Report,* pp. 185, 197.

39. Patterson to Eberstadt, Nov. 9, 1948, Baldwin to Eberstadt, Oct. 26, 1948, both in Eberstadt Papers; Vandenberg to Symington, Jan. 12, 1949, series 47.3, Office of the Director, Bureau of the Budget Records, in Record Group 51, NA.

40. Eberstadt Notes for ABC Headline Edition, Dec. 16, 1948; Eberstadt to Krock, Aug. 27 and Dec. 9, 1948, all in Eberstadt Papers; *Eisenhower Papers,* 10:234–35, 291–92, 553–40.

41. Eberstadt to Forrestal, Dec. 6, 1948, Eberstadt Papers; *Eisenhower Papers,* 10:430–31.

42. *Eisenhower Papers,* 10:462; Webb, quoted in *Lilienthal Journals,* 2:385–86; Jack K. McFall Oral History, 1970, HSTL; Eberstadt to Hoover, Mar. 31, 1949, Eberstadt Papers; Eberstadt to Eisenhower, Dec. 7, 1948, Eberstadt Papers.

43. Eberstadt to Duffield, Feb. 28, 1951, Eberstadt Papers; *Forrestal Diaries,* pp. 476–78.

44. *Eisenhower Papers,* 10:461–62, 483–84.

45. *Eisenhower Papers,* 10:401–402, 448–51, 462; Ayers Diary, entry Apr. 21, 1948, Ayers Papers; Donovan, *Tumultuous Years,* p. 60.

46. Eberstadt to Armstrong, Feb. 28, 1949; Eberstadt Memo to Forrestal, Feb. 16, 1949; Eberstadt to Forrestal, Feb. 17, 1949; Eberstadt to Duffield, Feb. 28, 1951; Armstrong to Eberstadt, Feb. 21, 1949, all in Eberstadt Papers; *Eisenhower Papers,* 10:465–66. Forrestal's Appointment Book and Telephone Log indicated that Forrestal visited South Carolina Mar. 11–12, 1949 (Forrestal Papers).

47. *Forrestal Diaries,* p. 552; Rogow, *James Forrestal,* pp. 314–15; Donovan, *Tumultuous Years,* pp. 60–61.

48. Eberstadt to Forrestal, Mar. 3, 1949; Forrestal to Eberstadt, Mar. 5, 1949 (last correspondence?), Eberstadt Papers.

49. Strauss, quoted in *Lilienthal Journals,* 2:506; *Eisenhower Papers,* 10:544–46; Krock to Eberstadt, Mar. 7, 1949, Eberstadt Papers; Souers recalled that despite the absence of his name in a hospital log, he visited Forrestal in his room on Apr. 13, 1949, to search for hidden microphones or bugs (Souers to Osborne, Aug. 19, 1963, Sidney Souers Papers, HSTL); Forrestal Appointment Book for Mar. 11–12, 1949, Forrestal Papers; for lack of home life, see Rogow, *James Forrestal;* Strauss, *Men and Decisions,* pp. 155–62.

50. Krock Private Memo on interview with Eberstadt, May 6, 1949, Krock Papers; Frederick Eberstadt, interview with author, Sept. 22, 1988; Rogow, *James Forrestal,* p. 5; Telephone calls to Eberstadt, Forrestal Appointment Book and Telephone Log for Mar., 1949, Forrestal Papers.

51. Eberstadt to Hoover, Mar. 30, 1949; Eberstadt to Dr. Raines, May 24, 1949, Eberstadt Papers; Krock Private Memo on Interview with Eberstadt, May 6, 1949, Krock Papers; see also Rogow, *James Forrestal.*

52. Frederick Eberstadt, interview with author, Sept. 22, 1988; Eberstadt to H. A. Smith, June 1, 1949, Eberstadt Papers; Patterson to Forrestal, Apr. 23, 1949, Patterson Papers; Adjutant General to Eberstadt, May 23, 1949, Eberstadt Papers.

CHAPTER 10

1. Symington to Tydings, May 10, 1949; Royall to Hoover, Apr. 27, 1949, both in Eberstadt Papers.

2. Eberstadt to Eisenhower, July 1, Oct. 22, 1949, Dec. 29, 1950, Jan. 6, 1951, Eberstadt Papers; *Eisenhower Papers,* 11:1142, 1146, 1148; Griffith, "Eisenhower and Corporate Commonwealth," pp. 87–122.

3. *Eisenhower Papers,* 11:1145–46, 888.

4. Eisenhower Luncheon List, 1950; Good Man lists, both in Eberstadt Papers.

5. Eberstadt to Krock, Mar. 4, July 1, Oct. 6, 1949, July 31, 1950; Eberstadt to Lawrence, Feb. 25, 1949; Baldwin to Eberstadt, Mar. 14, 1951; Eberstadt to Lippmann, May 8, 1958; Eberstadt to Henry Luce, Nov. 15, 1951, all in Eberstadt Papers; *Time,* Oct. 30, 1950, p. 30.

6. Laughlin to Eberstadt, June 23, 1949; Eberstadt to Leva, Jan. 23, 1964; Eberstadt to Laughlin, Nov. 17, 1949; Eberstadt to Dillon, June 24, 1949, all in Eberstadt Papers. See also Rogow, *James Forrestal;* Albion and Connery, *Forrestal and the Navy; Forrestal Diaries.*

7. Eberstadt to Patterson, June 24, 1949; Edwards to Bermingham, June 30, 1949; Eberstadt to Dillon, May 27, June 1, 1949; Eberstadt to Fennelly, June 26, 1962, Dec. 3, 1963, Nov. 23, 1964; Fennelly to Eberstadt, Aug. 3, 1964, all in Eberstadt Papers; Fennelly, *Memoirs of a Bureaucrat.*

8. Eberstadt's Statement to the Rockefeller Committee, Mar. 21, 1953, typescript, Eberstadt Papers.

9. Eberstadt to Admiral Radford, Mar. 16, May 6, 1953; Eberstadt to Hoover, Mar. 12 and 26, 1953; Radford to Eberstadt, Apr. 6, 1953; Eberstadt to Krock, May 7 and 8, 1953, all in Eberstadt Papers.

10. Foster to Eberstadt, June 22, 1953; Hittle to Eberstadt, Feb. 7, 1958, both in Eberstadt Papers.

11. Eberstadt to Hoover, May 6, 1958; Eberstadt to Nelson, Sept. 17, 1960; Eberstadt to Price, Oct. 14, 1960, all in Eberstadt Papers.

12. Eberstadt to "Jo" Forrestal, Jan. 8, 1959; Eberstadt to Leva, Nov. 14, 1960, Apr. 1, 1965, all in Eberstadt Papers; Perez and Willett, *Will to Win,* pp. 113, 126–27.

13. Eberstadt to Hoover, May 6, 1958; Eberstadt to Allen, Aug. 18, 1960; Eberstadt to McNamara, Dec. 11, 1967, all in Eberstadt Papers.

14. Baldwin to Christman, June 3, 1970, Christman Papers; see Hammond, *Organizing for Defense,* pp. 230–32. See also Hadley, *Straw Giant.*

15. Draft of Speech, 3/20/61, Eberstadt Papers.

16. Hensel's Speech, New Jersey Real Estate Appraisers Convention, Atlantic City, Nov. 20, 1945, Eberstadt Papers; Eberstadt interview with Christman, July 17, 1969, Princeton.

17. Donovan, *Tumultuous Years,* pp. 158–61; Hammond, "NSC-68: Prologue to Rearmament," in Schilling, Hammond, and Snyder, *Strategy, Politics, and Defense,* pp. 271–378.

18. Eberstadt to Krock, July 31, 1950, Eberstadt Papers.

19. Kinnard, *Secretary of Defense,* pp. 45–47, 201; Burch, Jr., *Elites in American History,* 3:143–45; Schmitter, "Century," in Pike and Stritch, eds., *New Corporatism,* pp. 101–11.

20. Kinnard, *Secretary of Defense,* pp. 77–120; Burch, Jr., *Elites in American History,* 3:209–10; Schmitter, "Century," in Pike and Stritch, eds., *New Corporatism,* p. 101.

21. Donovan, *Conflict and Crisis,* pp. 308–11; Burch, Jr., *Elites in American History,* 3:97, 164, 204, 219, 227–28; William R. Corson, *The Armies of Ignorance: The Rise of the American Intelligence Empire* (New York: Dial/James Wade, 1977), pp. 295–97, 469–71.

22. Borklund, *Department of Defense,* pp. 75–101.

23. Ibid., pp. 75–83; Cuff, "Eberstadt, NSRB," pp. 50–51; Hitch, in Hewes, *From Root to McNamara,* pp. 306–10. See also Hadley, *Straw Giant.*

24. Danhof, *Government Contracting,* p. 182; lists of top one hundred corporations, Pursell, ed., *Military-Industrial Complex,* pp. 320–21. See also, Barnet, *Roots of War.*

25. Boyd to Christman, May 20, 1970, Christman Papers. Barnet argues that the national security managers developed "the theater of crisis into a high art" (*Roots of War,* p. 267).

26. Draft speech of Mar. 20, 1961; Eberstadt to Leva, Aug. 13, 1947, both in Eberstadt Papers.

Bibliography

ARCHIVAL COLLECTIONS

George C. Marshall Library, Lexington, Virginia
- George C. Marshall Research File (photocopies)

Library of Congress, Washington, D.C.
- Ira C. Eaker Papers
- Merritt A. Edson Papers
- William F. Halsey Papers
- Ernest J. King Papers
- Frank Knox Papers
- Julius A. Krug Papers
- William D. Leahy Diaries
- John M. Palmer Papers
- Robert P. Patterson Papers
- William S. Sims Papers
- Carl Spaatz Papers

National Archives, Washington D.C.
- Record Group 51, Records of the Office of Management and Budget
- Record Group 80, Records of the Office of the Secretary of the Navy
  - Records of the Undersecretary and Secretary, 1941–1947
  - Minutes of the Top Policy Group Meetings
- Record Group 107, Records of the Office of the Secretary of War Patterson's Safe File
- Record Group 165, Records of the War Department General and Special Staffs
  - Plans and Operations Division, War Department, Joint Army-Navy Navy Board Numerical File
  - Records relating to War Department Reorganization and Armed Services Unification, 1939–47.
- Record of the Office of the Chief of Staff, 1903–47
  - Records relating to the Proposed Department of National Defense (Nelson File)
- Record Group 219, Joint Chief of Staff Records

Record Group 225, Records of the Joint Army-Navy Board Numerical File
Record Group 319, Records of the Army Staff
Records of the Office of the Chief of Staff, 1942–52
Record Group 340, Records of the Office of the Secretary of the Air Force
Records of the Air Board (Knerr File)
Public Information Division (Unification File)
Record Group 341, Records of Headquarters USAF
Naval Historical Division, Operational Archives, Washington, D.C.
General Board Records
Papers Relating to Secretary of the Navy James V. Forrestal
Papers of Admiral Harry E. Yarnell
Princeton University (Seeley Mudd Manuscript Library), Princeton, New Jersey
Bernard M. Baruch Papers
Edward M. Earle Papers
Ferdinand Eberstadt Papers
James V. Forrestal Papers
Arthur Krock Papers
Franklin D. Roosevelt Library, Hyde Park, N.Y.
President's Official File
President's Secretary's File
Samuel I. Rosenman Papers
Harold D. Smith Papers
Harry S. Truman Library, Independence, Mo.
Eben A. Ayers Papers
Clark M. Clifford Papers
George M. Elsey Papers
W. John Kenney Papers
Samuel I. Rosenman Papers
Harold Smith Papers
Sidney Souers Papers
Harry S. Truman Papers
President's Official File
President's Secretary's File
James E. Webb Papers
Yale University Library, New Haven, Conn.
Henry L. Stimson Diary (microfilm)

INTERVIEWS

Interviews with the author
Frederick Eberstadt, 1988

William Bradford Huie, 1984
Milton Katz, 1984
Princeton University (Seeley Mudd Manuscript Library)
Ferdinand Eberstadt. Interviews with Calvin L. Christman. Typed transcript, 1969.
Ferdinand Eberstadt. Interview. The John Foster Dulles Oral History Project, 1965.
Harry S. Truman Library, Oral History Interviews
John Abbot, 1970
Robert L. Dennison, 1972
George M. Elsey, 1974
John A. Kennedy, 1978
Marx Leva, 1972
Jack K. McFall, 1970
John L. Sullivan, 1974
James L. Sundquist, 1964
Harry H. Vaughan, 1964
U.S. Marine Corps Historical Center, Washington, D.C., Oral History Unit, Transcripts of Interviews
Victor H. Krulak
Alexander A. Vandegrift
U.S. Naval Institute, Oral History Project, Annapolis, Md.
Hanson W. Baldwin, 1975
William R. Smedberg, 1979

BOOKS, ARTICLES, AND DISSERTATIONS

Abell, Tyler, ed. *Drew Pearson: Diaries, 1949–1959*. New York: Holt, Rinehart and Winston, 1974.

Acheson, Dean. *Present at the Creation: My Years in the State Department*. New York: W. W. Norton, 1969.

Albion, Robert Greenhalgh, and Robert Howe Connery. *Forrestal and the Navy*. New York: Columbia University Press, 1962.

Allen, James, ed. *Democracy and Finance: The Addresses and Public Statements of William O. Douglas*. 1940. Reprint. Port Washington, N.Y.: Kennikat, 1968.

Ambrose, Stephen E., ed. *Institutions in Modern America: Innovation in Structure and Process*. Baltimore: Johns Hopkins University Press, 1967.

Arnold, Henry H. *Global Mission*. New York: Harper and Brothers, 1949.

Barnet, Richard J. *Roots of War: The Men and Institutions behind U.S. Foreign Policy*. Baltimore: Penguin, 1973.

Baruch, Bernard M. *Baruch: The Public Years*. New York: Holt, Rinehart and Winston, 1960.

Berle, Adolf. *Navigating the Rapids, 1918–1971.* New York: Harcourt Brace Jovanovich, 1973.

Berman, Larry. *The Office of Management and Budget and the Presidency, 1921– 1979.* Princeton, N.J.: Princeton University Press, 1979.

Bland, Larry I., ed. *The Papers of George Catlett Marshall.* 2 vols. Baltimore: Johns Hopkins University Press, 1981, 1986.

Block, Fred L. *The Origins of International Economic Disorder: A Study of United States International Monetary Policy from World War II to the Present.* Berkeley: University of California Press, 1977.

Blum, John Morton. *V Was for Victory: Politics and American Culture During World War II.* New York: Harcourt Brace Jovanovich, 1976.

Blum, John Morton, ed. *The Price of Vision: The Diary of Henry A. Wallace, 1942–1946.* Boston: Houghton Mifflin, 1973.

Borklund, C. W. *The Department of Defense.* New York: Frederick A. Praeger, 1968.

Borklund Carl W. *Men of the Pentagon: From Forrestal to McNamara.* New York: Praeger, 1966.

Boulding, Kenneth E. *The Organizational Revolution: A Study in the Ethics of Economic Organization.* 1953. Reprint. Chicago: Quadrangle, 1968.

Bowen, Ralph H. *German Theories of the Corporative State with Special Reference to the Period 1870–1919.* New York: Whittlesey House–McGraw-Hill, 1947.

Breen, William J. *Uncle Sam at Home: Civilian Mobilization, Wartime Federalism, and the Council of National Defense, 1917–1919.* Westport, Conn.: Greenwood, 1984.

Brown, Harold. *Thinking about National Security: Defense and Foreign Policy in a Dangerous World.* Boulder, Colo.: Westview Press, 1983.

Brownlow, Louis. *A Passion for Anonymity: The Autobiography of Louis Brownlow, Second Half.* Chicago: University of Chicago Press, 1958.

Brune, Lester H. *The Origins of American National Security Policy: Sea Power, Air Power and Foreign Policy, 1900–1941.* Manhattan: MA/AH Publishing, Kansas State University, 1981.

Buell, Thomas B. *Master of Sea Power: A Biography of Fleet Admiral Ernest J. King.* Boston: Little, Brown, 1980.

Burch, Philip H., Jr. *Elites in American History.* 3 vols. New York: Holmes and Meier, 1981.

Byrnes, James F. *Speaking Frankly.* New York: Harper and Brothers, 1947.

Cantril, Hadley. *Public Opinion, 1935–1946.* Princeton, N.J.: Princeton University Press, 1951.

Caraley, Demetrios. *The Politics of Military Unification: A Study of Conflict and the Policy Process.* New York: Columbia University Press, 1966.

Catton, Bruce. *The War Lords of Washington.* New York: Harcourt, Brace, 1948.

Challener, Richard D., and Gordon B. Turner, eds. *National Security in the Nuclear Age: Basic Facts and Theories.* New York: Frederick A. Praeger, 1961.

Chambers, John Whiteclay, II. *To Raise an Army: The Draft Comes to Modern America.* New York: Free Press, 1987.

Chandler, Alfred D., Jr. *The Visible Hand: The Managerial Revolution in American Business.* Cambridge, Mass.: Belknap Press of Harvard University Press, 1977.

Chandler, Alfred D., Jr., and Louis Galambos, eds. *The Papers of Dwight David Eisenhower.* 11 vols. Baltimore: Johns Hopkins University Press, 1970–84.

Christman, Calvin Lee. "Ferdinand Eberstadt and Economic Mobilization for War, 1941–1943." Ph.D. diss., Ohio State University, 1971.

———. "Charles A. Beard, Ferdinand Eberstadt, and America's Postwar Security." *Mid-America* 54 (July 1972): 187–94.

———. "Donald Nelson and the Army." *Military Affairs* 37 (1973): 81–83.

Clarkson, Grosvenor B. *Industrial America in the World War: The Strategy behind the Line, 1917–1918.* Boston: Houghton Mifflin, 1923.

Cleveland, Harold Van B., and Thomas F. Huertas. *Citibank, 1812–1970.* Cambridge, Mass.: Harvard University Press, 1985.

Clifford, John Garry. *The Citizen Soldiers: The Plattsburg Training Camp Movement, 1913–1920.* Lexington: University Press of Kentucky, 1972.

Cline, Ray S. *Washington Command Post: The Operations Division.* Washington, D.C.: Office of the Chief of Military History, U.S. Army, 1951.

Cline, Ray S., and Maurice Matloff. "Development of War Department Views on Unification." *Military Affairs* 13 (Summer, 1949): 65–74.

Coffey, Thomas, M. *Hap: The Story of the U.S. Air Force and the Man Who Built It: General Henry "Hap" Arnold.* New York: Viking, 1982.

Coles, Harry L., ed. *Total War and Cold War: Problems in Civilian Control of the Military.* Columbus: Ohio State University Press, 1962.

Coletta, Paolo E. *The United States Navy and Defense Unification, 1947–1953.* Newark: University of Delaware Press, 1981.

Coletta, Paolo E., ed. *American Secretaries of the Navy.* 2 vols. Annapolis, Md.: Naval Institute Press, 1980.

Collins, Robert M. *The Business Response to Keynes, 1929–1964.* New York: Columbia University Press, 1981.

———. "Positive Business Responses to the New Deal: The Roots of the Committee for Economic Development, 1933–1942." *Business History Review* 12 (Autumn, 1978): 369–91.

Connery, Robert H. *The Navy and the Industrial Mobilization in World War II.* Princeton, N.J.: Princeton University Press, 1951.

Cooling, B. Franklin. "The Formative Years of the Naval-Industrial Complex: Their Meaning for Studies of Institutions Today." *Naval War College Review* 27 (April–March, 1975): 53–61.

Costigliola, Frank. "The United States and the Reconstruction of Germany in the 1920s." *Business History Review* 50 (Winter, 1976): 477–502.

Cotton, Joseph P., Jr. "Recent New York Legislation upon Workmen's Compensation." *Annals of the American Academy of Political Social Science* 38 (1911): 230–37.

Critchlow, Donald T. *The Brookings Institution, 1916–1952: Expertise and the Public Interest in a Democratic Society.* DeKalb: Northern Illinois University Press, 1985.

Cronin, Thomas E., and Sanford D. Greenberg. *The Presidential Advisory System.* New York: Harper and Row, 1969.

Cronon, E. David, ed. *The Cabinet Diaries of Josephus Daniels, 1913–1921.* Lincoln: University of Nebraska Press, 1963.

———. *The Political Thought of Woodrow Wilson.* Indianapolis: Bobbs-Merrill, 1965.

Cuff, Robert D. "Ferdinand Eberstadt, the National Security Resources Board, and the Search for Integrated Mobilization Planning, 1947–1948." *Public Historian* 7 (Fall, 1985): 37–52.

———. *The War Industries Board: Business-Government Relations during World War I.* Baltimore: Johns Hopkins University Press, 1973.

———. "Bernard Baruch: Symbol and Myth in Industrial Mobilization." *Business History Review* 43 (Summer, 1969): 115–33.

———. "From the Controlled Materials Plan to the Defense Materials System, 1942–1953." *Military Affairs* 51 (January, 1987): 1–6.

———. "Herbert Hoover, the Ideology of Voluntarism, and War Organization during the Great War." *Journal of American History* 64 (September, 1977): 358–72.

———. "An Organizational Perspective on the Military-Industrial Complex." *Business History Review* 52 (Summer, 1978): 250–67.

Danhof, Clarence H. *Government Contracting and Technological Change.* Washington, D.C.: Brookings Institution, 1968.

Daniels, Jonathan. *Frontier on the Potomac.* New York: Macmillan, 1946.

———. *White House Witness, 1942–1945.* Garden City, N.Y.: Doubleday, 1975.

Davis, Vincent. *The Admirals Lobby.* Chapel Hill: University of North Carolina Press, 1967.

———. *Postwar Defense Policy and the U.S. Navy, 1943–1946.* Chapel Hill: University of North Carolina Press, 1966.

Dawes, Charles G. *A Journal of the Great War.* 2 vols. Boston: Houghton Mifflin, 1921.

Dewey, Thomas E. *The Case against the New Deal.* New York: Harper and Brothers, 1940.

Domhoff, G. William. *Who Rules America?* Englewood Cliffs, N.J.: Prentice-Hall, 1967.

Donovan, Robert J. *Tumultuous Years: The Presidency of Harry S. Truman, 1949–1953.* New York: W. W. Norton, 1982.

———. *Conflict and Crisis: The Presidency of Harry S. Truman, 1945–1948*. New York: W. W. Norton, 1977.

Eberstadt, Ferdinand. "A Friend's Recollection of Dr. Carl Melchior." In *Dr. Carl Melchior: Ein Buch des Gedenkens und der Freundschaft*, pp. 121–22. Tübingen: Paul Siebeck, 1967.

———. "What Did We Learn about Controls in World War II?" *Current Business Studies* 7 (December, 1950): 3–9.

Elliott, William Yandell. *The Need for Constitutional Reform: A Program for National Security*. New York: Whittlesey House, 1935.

Emmerich, Herbert. *Federal Organization and Administrative Management*. University: University of Alabama Press, 1971.

Epstein, Edwin M. *The Corporation in American Politics*. Englewood Cliffs, N.J.: Prentice-Hall, 1969.

Fanton, Jonathan Foster. "Robert A. Lovett: The War Years." Ph.D. diss., Yale University, 1978.

Fausold, Martin L., ed. *The Hoover Presidency: A Reappraisal*. Albany: State University of New York Press, 1974.

Fennelly, John F. *Memoirs of a Bureaucrat: A Personal Story of the War Production Board*. Chicago: October House, 1965.

Ferrell, Robert H., ed. *The Eisenhower Diaries*. New York: W. W. Norton, 1981.

———. *Off the Record: The Private Papers of Harry S. Truman*. New York: Harper and Row, 1980.

Fetter, Frank A. "Subsidizing of Private Charities." *American Journal of Sociology* 7 (November, 1901): 359–85.

———. "State Supervision and Administration." *Survey* 22 (June, 1909): 455–57.

Foulois, Benjamin D., and C. V. Glines. *From the Wright Brothers to the Astronauts: The Memoirs of Major General Benjamin D. Foulois*. New York: McGraw-Hill, 1968.

Fox, Daniel M. *The Discovery of Abundance: Simon N. Patten and the Transformation of Social Theory*. Ithaca, N.Y.: Cornell University Press, 1967.

Fox, Richard Wightman. *Reinhold Niebuhr: A Biography*. New York: Pantheon, 1985.

Freedman, Max, ed. *Roosevelt and Frankfurter: Their Correspondence, 1928–45*. Boston: Little, Brown, 1967.

Galambos, Louis. *Competition & Cooperation: The Emergence of a National Trade Association*. Baltimore: Johns Hopkins University Press, 1966.

———. "The Emerging Organizational Synthesis In Modern American History." *Business History Review* 44 (Autumn, 1970): 279–90.

———. "Technology, Political Economy, and Professionalization: Central Themes of the Organizational Synthesis." *Business History Review* 57 (Winter, 1983): 471–93.

Galambos, Louis, ed. *The New American State: Bureaucracies and Policies since World War II*. Baltimore: Johns Hopkins University Press, 1987.

Gardner, Lloyd C. *Architects of Illusion: Men and Ideas in American Foreign Policy, 1941–1949*. Chicago: Quadrangle, 1970.

Gaus, John Merriman. *Reflections on Public Administration*. Montgomery: University of Alabama Press, 1947.

Greenbaum, Edward S. *A Lawyer's Job: In Court, in the Army, in the Office*. New York: Harcourt, Brace and World, 1967.

Griffis, Stanton. *Lying in State*. Garden City, N.Y.: Doubleday, 1952.

Griffith, Robert. "Dwight Eisenhower and the Corporate Commonwealth," *American Historical Review* 87 (February, 1982): 87–122.

Gulick, Luther. *Administrative Reflections from World War II*. University: University of Alabama Press, 1948.

Hadley, Arthur T. *The Straw Giant: America's Armed Forces: Triumphs and Failures*. New York: Avon, 1987.

Hammond, Paul Y. *Organizing for Defense: The American Military Establishment in the Twentieth Century*. Princeton, N.J.: Princeton University Press, 1961.

Harriman, Roland. *I Reminisce*. New York: Doubleday, 1975.

Harrison, Reginald J. *Pluralism and Corporatism: The Political Evolution of Modern Democracies*. London: George Allen and Unwin, 1980.

Hartmann, Susan M. *Truman and the 80th Congress*. Columbia: University of Missouri Press, 1971.

Hassler, Warren W., Jr. *With Shield and Sword: American Military Affairs, Colonial Times to the Present*. Ames: Iowa State University Press, 1982.

Hawley, Ellis W. "The Discovery and Study of a 'Corporate Liberalism.'" *Business History Review* 52 (Autumn, 1978): 309–30.

———. *The Great War and the Search for a Modern Order: A History of the American People and Their Institutions, 1917–1933*. New York: St. Martin's Press, 1979.

———. "Herbert Hoover, the Commerce Secretariat, and the Vision of an 'Associate State.'" *Journal of American History* 61 (June, 1974): 116–40.

———. *The New Deal and the Problem of Monopoly: A Study in Economic Ambivalence*. Princeton: N.J.: Princeton University Press, 1966.

Hayes, Grace Person. *The History of the Joint Chiefs of Staff in World War II: The War Against Japan*. Annapolis, Md.: Naval Institute Press, 1982.

Herbst, Jurgen. *The German Historical School in American Scholarship: A Study in the Transfer of Culture*. Ithaca, N.Y.: Cornell University Press, 1965.

Herring, E. Pendleton, ed. *Civil-Military Relations: Bibliographical Notes on Administrative Problems of Civilian Mobilization*. Chicago: Public Administration Service, 1940.

———. *Group Representation before Congress*. Baltimore: Johns Hopkins University Press, 1929.

———. *The Impact of War: Our Democracy under Arms*. New York: Farrar and Rinehart, 1941.

———. *Public Administration and the Public Interest*. 1936. Reprint. New York: Russell and Russell, 1967.

Hewes, James E., Jr. *From Root to McNamara: Army Organization and Administration, 1900–1963*. Washington, D.C.: Center of Military History, U.S. Army, 1975.

Himmelberg, Robert F. *The Origins of the National Recovery Administration: Business, Government, and the Trade Association Issue, 1921–1933*. New York: Fordham University Press, 1976.

Hobbs, Joseph Patrick. *Dear General: Eisenhower's Wartime Letters to Marshall*. Baltimore: Johns Hopkins University Press, 1971.

Hogan, Michael J. *Informal Entente: The Private Structure of Cooperation in Anglo-American Economic Diplomacy, 1918–1928*. Columbia: University of Missouri Press, 1977.

———. *The Marshall Plan: America, Britain, and the Reconstruction of Western Europe, 1947–1952*. Cambridge: Cambridge University Press, 1987.

Holley, I. B., Jr. *General John M. Palmer, Citizen Soldiers, and the Army of a Democracy*. Westport, Conn.: Greenwood, 1982.

Hoover, Herbert. *Herbert Hoover: An American Epic*. 4 vols. Chicago: Henry Regnery, 1959–64.

Hoover Commission Report on Organization of the Executive Branch of the Government. New York: McGraw-Hill, 1949.

Horowitz, David, ed. *Corporations and the Cold War*. New York: Monthly Review Press, 1969.

Huie, William Bradford. *The Case against the Admirals: Why We Must Have a Unified Command*. New York: E. P. Dutton, 1946.

———. "Untold Facts in the Forrestal Case." *New American Mercury* 71 (December, 1950): 643–52.

Huntington, Samuel P., ed. *Changing Patterns of Military Politics*. Glencoe, Ill.: The Free Press, 1962.

———. *The Common Defense: Strategic Programs in National Politics*. New York: Columbia University Press, 1961.

Huthmacher, J. Joseph, and Warren I. Susman, eds. *Herbert Hoover and the Crisis of American Capitalism*. Cambridge, Mass.: Schenkman, 1973.

Huzar, Elias. "Notes on the Unification Controversy." *Public Administration Review* (Autumn, 1946): 297–314.

Hyneman, Charles S. *The Study of Politics: The Present State of American Political Science*. Urbana: University of Illinois Press, 1959.

Isaacson, Walter, and Evan Thomas. *The Wise Men: Six Friends and the World They Made*. New York: Simon and Schuster, 1986.

Israel, Jerry, ed. *Building the Organizational Society: Essays on Associational Activities in Modern America*. New York: The Free Press, 1972.

Janeway, Eliot. *The Economics of Crisis: War, Politics, and the Dollar.* New York: Weybright and Talley, 1968.

———. *The Struggle for Survival: A Chronicle of Economic Mobilization in World War II*. New Haven, Conn.: Yale University Press, 1951.

Jessup, John K., ed. *The Ideas of Henry Luce*. New York: Atheneum, 1969.

Johnson, Franklyn A. *Defense by Committee: The British Committee of Imperial Defense, 1885–1959*. London: Oxford University Press, 1960.

Jurika, Stephen B., Jr., ed. *From Pearl Harbor to Vietnam: The Memoirs of Admiral Arthur W. Radford*. Stanford: Hoover Institution Press, 1980.

Kahn, E. J., Jr. *The World of Swope*. New York: Simon and Schuster, 1965.

Karl, Barry Dean. *Charles E. Merriam and the Study of Politics*. Chicago: University of Chicago Press, 1974.

———. *Executive Reorganization and Reform in the New Deal: The Genesis of Administrative Management, 1900–1939*. Cambridge, Mass.: Harvard University Press, 1963.

———. *The Uneasy State: The United States from 1915 to 1945*. University of Chicago Press, 1983.

Kaufman, Burton I. *Efficiency and Expansion: Foreign Trade Organization in the Wilson Administration, 1913–1921*. Westport, Conn.: Greenwood Press, 1974.

Keiser, Gordon W. *The U.S. Marine Corps and Defense Unification, 1944–1947: The Politics of Survival*. Washington, D.C.: National Defense University Press, 1982.

Kennan, George P. *Memoirs, 1925–1950*. New York: Bantam, 1969.

Kennedy, David M. *Over Here: The First World War and American Society*. Oxford: Oxford University Press, 1980.

Kennedy, Susan Estabrook. *The Banking Crisis of 1933*. Lexington: University of Kentucky Press, 1973.

Kinnard, Douglas. *The Secretary of Defense*. Lexington: University Press of Kentucky, 1980.

Kiplinger, W. M. *Washington Is Like That*. New York: Harper and Brothers, 1942.

Kirkendall, Richard S., ed. *The Truman Period as a Research Field: A Reappraisal, 1972*. Columbia: University of Missouri Press, 1974.

Koistinen, Paul A. C. "The 'Industrial-Military Complex' in Historical Perspective: The Interwar Years," *Journal of American History* 56 (March, 1970): 819–39.

Korb, Lawrence J. *The Joint Chiefs of Staff: The First Twenty-Five Years*. Bloomington: Indiana University Press, 1976.

Krock, Arthur. *Memoirs: Sixty Years on the Firing Line*. New York: Funk and Wagnalls, 1968.

Krooss, Herman E. *Executive Opinion: What Business Leaders Said and Thought on Economic Issues, 1920s–1960s.* Garden City, N.Y.: Doubleday, 1970.

Lane, Jack C. *Armed Progressive: General Leonard Wood.* San Rafael, Calif.: Presidio Press, 1978.

Lash, Joseph P. *From the Diaries of Felix Frankfurter with a Biographical Essay and Notes.* New York: W. W. Norton, 1975.

Leahy, William D. *I Was There: The Personal Story of the Chief of Staff to Presidents Roosevelt and Truman Based on His Notes and Diaries Made at the Time.* New York: Whittlesey House–McGraw-Hill, 1950.

Lee, David D. "Herbert Hoover and the Development of Commercial Aviation, 1921–1926." *Business History Review* 58 (Spring, 1984): 78–102.

Leffler, Melvyn P. "The American Conception of National Security and the Beginnings of the Cold War, 1945–1948." *American Historical Review* 89 (April, 1984): 346–81.

Legere, Maj. Lawrence Joseph. "Unification of the Armed Forces." Ph.D. diss., Harvard University, 1950.

Leslie, Frank P. *James Forrestal: Princeton Class of 1915.* N.p., 1951.

Lilienthal, David E. *The Journals of David E. Lilienthal: The Atomic Energy Years, 1945–1950.* New York: Harper and Row, 1964.

Link, Arthur S., ed. *The Papers of Woodrow Wilson.* Vol. 27. Princeton, N.J.: Princeton University Press, 1978.

Lippman, Walter. *Drift and Mastery: An Attempt to Diagnose the Current Unrest.* 1914. Reprint. Englewood Cliffs, N.J.: Prentice-Hall, 1961.

———. *U.S. Foreign Policy: Shield of the Republic.* Boston: Little, Brown, 1943.

Livermore, Seward W. *Woodrow Wilson and the War Congress, 1916–18.* Seattle: University of Washington Press, 1968.

Livezey, William E. *Mahan on Sea Power.* Norman: University of Oklahoma Press, 1981.

Lyons, Gene M., Louis Morton, and John W. Masland. *Schools for Strategy: Education and Research in National Security Affairs.* New York: Frederick A. Praeger, 1965.

———. *The Uneasy Partnership: Social Science and the Federal Government in the Twentieth Century.* New York: Russell Sage Foundation, 1969.

McClendon, R. Earl. *The Question of Autonomy for the United States Air Arm, 1907–1945.* Maxwell AFB, Ala.: Air University Documentary Research Division, 1954.

McCormick, Thomas J. "Drift or Mastery? A Corporatist Synthesis for American Diplomatic History." *Reviews in American History* 10 (December, 1982): 318–30.

McQuaid, Kim. "Competition, Cartellization, and the Corporate Ethic: General Electric's Leadership During the New Deal, 1933–40." *American Journal of Economics and Sociology* 36 (October, 1977): 417–28.

———. "Corporate Liberalism in the American Business Community, 1920–1940." *Business History Review* 52 (Autumn, 1978): 342–68.

———. "Young, Swope and General Electric's 'New Capitalism': A Study in Corporate Liberalism, 1920–33." *American Journal of Economics and Sociology* 36 (July, 1977): 323–34.

Maier, Charles S. *Recasting Bourgeois Europe: Stabilization in France, Germany, and Italy in the Decade after World War I.* Princeton, N.J.: Princeton University Press, 1975.

Mann, Dean E., and Jameson W. Doig. *The Assistant Secretaries: Problems and Processes of Appointment.* Washington, D.C.: The Brookings Institution, 1965.

Mathey, Dean. *Fifty Years of Wall Street with Anecdotiana.* Princeton, N.J.: privately printed, 1966.

May, Ernest R., ed. *The Ultimate Decision: The President as Commander in Chief.* New York: George Braziller, 1960.

Mayer, Martin. *Emory Buckner.* New York: Harper and Row, 1968.

Messer, Robert L. *The End of an Alliance: Byrnes, Roosevelt, Truman, and the Origins of the Cold War.* Chapel Hill: University of North Carolina Press, 1982.

Meyer, Richard Hemmig. *Bankers' Diplomacy: Monetary Stabilization in the Twenties.* New York: Columbia University Press, 1970.

Millett, Allan R., and Peter Maslowski. *For the Common Defense: A Military History of the United States of America.* New York: The Free Press, 1984.

Millett, John D. *The Army Service Forces: The Organization and Role of the Army Service Forces.* Washington, D.C.: Office of the Chief of Military History, 1954.

Millis, Walter, ed. *American Military Thought.* Indianapolis: Bobbs-Merrill Company, 1966.

Millis, Walter, and E. S. Duffield, eds. *The Forrestal Diaries.* New York: Viking, 1951.

Millis, Walter, Harvey C. Mansfield, and Harold Stein, eds. *Arms and the State: Civil-Military Elements in National Policy.* New York: Twentieth Century Fund, 1958.

Mitchell, Wesley C. *Types of Economic Theory from Mercantilism to Institutionalism.* 2 vols. New York: Augustus M. Kelley, 1969.

Moe, Ronald C. *The Hoover Commissions Revisited.* Boulder, Colo.: Westview Press, 1982.

Morison, Elting E. *Turmoil and Tradition: A Study of the Life and Times of Henry L. Stimson.* Boston: Houghton Mifflin, 1960.

Morse, Philip M. *In at the Beginnings: A Physicist's Life.* Cambridge, Mass.: MIT Press, 1977.

Mrozek, Donald John. "Peace through Strength: Strategic Air Power and the Mobilization of the United States for the Pursuit of Foreign Policy, 1945–1955." Ph.D. diss., Rutgers University, 1972.

———. "The Truman Administration and Enlistment of the Aviation Industry in Postwar Defense." *Business History Review* 48 (Spring, 1974): 72–94.

Mulder, John M. *Woodrow Wilson: The Years of Preparation.* Princeton, N.J.: Princeton University Press, 1978.

Nelson, Daniel. *Frederick W. Taylor and the Rise of Scientific Management.* Madison: University of Wisconsin Press, 1980.

Nelson, Donald M. *Arsenal of Democracy: The Story of American War Production.* New York: Harcourt, Brace, 1946.

Nelson, Otto L., Jr. *National Security and the General Staff.* Washington, D.C.: Infantry Journal Press, 1946.

Newton, James Y., "Eberstadt: The Man behind the Big Brass." *American Magazine* (May, 1949): 30, 87.

Nicolson, Harold. *Dwight Morrow.* New York: Harcourt, Brace, 1935.

Novick, David, Melvin Anshen, and W. C. Truppner. *Wartime Production Controls.* New York: Columbia University Press, 1949.

O'Connor, Raymond G., ed. *American Defense Policy in Perspective from Colonial Times to the Present.* New York: John Wiley and Sons, 1965.

Osgood, Charles G. *Lights in Nassau Hall: A Book of the Bicentennial, Princeton, 1746–1946.* Princeton, N.J.: Princeton University Press, 1951.

Palmer, John McAuley. *America in Arms: The Experience of the United States with Military Organization.* New Haven, Conn.: Yale University Press, 1941.

———. *Statesmanship or War.* Garden City, N.Y.: Doubleday and Page, 1927.

Parrish, Michael E. *Securities Regulation and the New Deal.* New Haven, Conn.: Yale University Press, 1970.

Patterson, James T. *Congressional Conservatism and the New Deal: The Growth of the Conservative Coalition in Congress, 1933–1939.* Lexington: University of Kentucky Press, 1967.

Pearlman, Michael. *To Make Democracy Safe for America: Patricians and Preparedness in the Progressive Era.* Urbana: University of Illinois Press, 1984.

Pemberton, William E. *Bureaucratic Politics: Executive Reorganization during the Truman Administration.* Columbia: University of Missouri Press, 1979.

Perez, Robert C., and Edward F. Willett. *The Will to Win: A Biography of Ferdinand Eberstadt.* Westport, Conn.: Greenwood Press, 1989.

Perkins, Edwin J., ed. *Men and Organizations: The American Economy in the Twentieth Century.* New York: G. P. Putnam's Sons, 1977.

Pfau, Richard. *No Sacrifice Too Great: The Life of Lewis L. Strauss.* Charlottesville: University Press of Virginia, 1984.

Pike, Frederick B., and Thomas Stritch, eds. *The New Corporatism: Social-Political Structures in the Iberian World.* Notre Dame, Ind.: University of Notre Dame Press, 1974.

Pilat, Oliver. *Drew Pearson: An Unauthorized Biography.* New York: Pocket Books, 1973.

Poen, Monte M., ed. *Strictly Personal and Confidential: The Letters Harry Truman Never Mailed*. Boston: Little, Brown, 1982.

Pogue, Forrest C. *George C. Marshall: Ordeal and Hope, 1939–1942*. New York: Viking, 1965–66.

———. *George C. Marshall: Organizer of Victory, 1943–1945*. New York: Viking, 1973.

Polenberg, Richard. *Reorganizing Roosevelt's Government: The Controversy over Executive Reorganization, 1936–1939*. Cambridge, Mass.: Harvard University Press, 1966.

Public Papers of the Presidents of the United States: Harry S. Truman, Containing the Public Messages, Speeches, and Statements of the President, April 12 to December 31, 1945. Washington, D.C.: Government Printing Office, 1961.

Pursell, Carroll W., Jr. *The Military-Industrial Complex*. New York: Harper and Row, 1972.

Ransom, Harry Howe. *Central Intelligence and National Security*. Cambridge, Mass.: Harvard University Press, 1958.

Raymond, Jack. *Power at the Pentagon*. New York: Harper and Row, 1964.

Rearden, Steven L. *History of the Office of the Secretary of Defense: The Formative Years, 1947–1950*. Washington, D.C.: Office of the Secretary of Defense, 1984.

Redfield, William C. *The New Industrial Day*. New York: Century, 1913.

Riddle, Donald H. *The Truman Committee: A Study in Congressional Responsibility*. New Brunswick, N.J.: Rutgers University Press, 1964.

Ridgway, Matthew B. *Soldier: The Memoirs of Matthew B. Ridgway*. New York: Harpers, 1956.

Ries, John C. *The Management of Defense: Organization and Control of the U.S. Armed Services*. Baltimore: Johns Hopkins University Press, 1964.

Rogow, Arnold A. *James Forrestal: A Study of Personality, Politics, and Policy*. New York: Macmillan, 1963.

Root, Elihu. *The Military and Colonial Policy of the United States: Addresses and Reports*. 1916. Reprint. New York: AMS Press, 1970.

Sander, Alfred D. "Truman and the National Security Council, 1945–1947." *Journal of American History* 59 (September, 1972): 369–88.

Schelling, Thomas C. *The Strategy of Conflict*. New York: Oxford University Press, 1963.

Schilling, Warner, Paul Y. Hammond, and Glenn H. Snyder. *Strategy, Politics, and Defense Budgets*. New York: Columbia University Press, 1962.

Schmitter, Philippe C., and Gerhard Lehmbruch. *Trends toward Corporatist Intermediation*. Beverly Hills, Calif.: Sage Publications, 1979.

Schulzinger, Robert D. *The Wise Men of Foreign Affairs: The History of the Council on Foreign Relations*. New York: Columbia University Press, 1984.

Schwarz, Jordan A. *Liberal: Adolf A. Berle and the Vision of an American Era*. New York: The Free Press, 1987.

———. *The Speculator: Bernard M. Baruch in Washington, 1917–1965.* Chapel Hill: University of North Carolina Press, 1981.

Seager, Robert, II. *Alfred Thayer Mahan: The Man and His Letters.* Annapolis, Md.: Naval Institute Press, 1977.

Sherry, Michael S. *Preparing for the Next War: American Plans for Postwar Defense, 1941–45.* New Haven, Conn.: Yale University Press, 1977.

Sherwood, Robert E. *Roosevelt and Hopkins: An Intimate History.* New York: Harper and Brothers, 1950.

Shonfield, Andrew. *Modern Capitalism: The Changing Balance of Public and Private Power.* London: Oxford University Press, 1965.

Skowronek, Stephen. *Building a New American State: The Expansion of National Administrative Capacities, 1877–1920.* Cambridge: Cambridge University Press, 1982.

Sloan, Alfred P., Jr. *My Years with General Motors.* Garden City, N.Y.: Doubleday, 1964.

Smith, Bruce L. R. *The Rand Corporation: Case Study of a Nonprofit Advisory Corporation.* Cambridge, Mass.: Harvard University Press, 1966.

Smith, Jean Edward, ed. *The Papers of General Lucius D. Clay: Germany, 1945–49.* 2 vols. Bloomington: Indiana University Press, 1974.

Smith, Perry McCoy. *The Air Force Plans for Peace, 1943–1945.* Baltimore: Johns Hopkins University Press, 1970.

Smith, R. Elberton. *The Army and Economic Mobilization.* Washington, D.C.: Office of the Chief of Military History, 1959.

Sprout, Harold, and Margaret Sprout, eds. *The Foundations of National Power.* New York: D. Van Nostrand, 1951.

Stanley, Timothy W. *American Defense and National Security.* Washington, D.C.: Public Affairs Press, 1956.

Stein, Harold, ed. *American Civil-Military Decisions: A Book of Case Studies.* Birmingham: University of Alabama Press, 1963.

Stimson, Henry L., and McGeorge Bundy. *On Active Service in Peace and War.* New York: Harper, 1947–48.

Strauss, Lewis L. *Men and Decisions.* Garden City, N.Y.: Doubleday, 1962.

Swaine, Robert T. *The Cravath Firm of Its Predecessors, 1819–1948.* New York: Ad Press, 1948.

Tarr, Curtis William. "Unification of America's Armed Forces: A Century and a Half of Conflict, 1798–1947." Ph.D. diss., Stanford University, 1961.

Troy, Thomas F. *Donovan and the CIA: A History of the Establishment of the Central Intelligence Agency.* Central Intelligence Agency, Center for the Study of Intelligence, Washington, D.C., 1981.

Truman, Harry S. *Map Room Messages of President Truman, 1945–1946.* Microfilm. Frederick, Md.: University Publications of America, 1980.

Truman, Margaret. *Harry S. Truman.* New York: Morrow, 1973.

U.S. Congress, House. Committee on Expenditures in the Executive Departments. *National Security Act of 1947, Hearings on H.R. 2319.* 80th Cong., 1st sess., 1947 (Hoffman Hearings).

———. Committee on Expenditures in the Executive Departments. *Proposing the Establishment of a Department of National Defense, Hearings on H.R. 4742 and H.R. 7012.* 72d Cong., 1st sess., 1932.

———. Committee on Military Affairs. *Department of Defense and Unification of Air Service, Hearings.* 69th Cong., 1st sess., 1926.

———. Select Committee on Post-War Military Policy (Woodrum Committee). *Proposal to Establish a Single Department of Armed Forces, Hearings.* 78th Cong., 2d sess., 1944.

U.S. Congress. Senate. Committee on Armed Services. *National Security Act of 1947, Hearings on S. 758.* 80th Cong., 1st sess., 1947 (Gurney Hearings).

———. Military Affairs Committee. *Department of Armed Forces, Hearings on S. 84 and S. 1482.* 79th Cong., 1st sess., 1945 (Thomas Hearings).

———. Committee on Naval Affairs. *Unification of the Armed Forces, Hearings on S. 2044: A Bill to Promote the Common Defense by Unifying the Departments and Agencies of the Government Relating to the Common Defense,* 79th Cong., 2d sess., 1946 (Walsh Hearings).

———. Committee on Naval Affairs. *Report to Honorable James Forrestal, Secretary of the Navy on Unification of the War and Navy Departments and Postwar Organization for National Security, October 22, 1945.* Senate Committee Print, 79th Cong., 1st sess., 1945 (Eberstadt Report).

U.S. Joint Army and Navy Board. *Joint Action of the Army and the Navy, Prepared by the Joint Board 1927.* Washington, D.C.: U.S. Government Printing Office, 1927.

Uselding, Paul, ed. *Research in Economic History.* Greenwich, Conn.: JAI Press, 1976.

Vandenberg, Arthur H., Jr., ed. *The Private Papers of Senator Vandenberg.* Boston: Houghton Mifflin, 1952.

Waldo, Dwight. *The Administrative State: A Study of the Political Theory of American Public Administration.* New York: Ronald Press, 1948.

Weber, Max. *The Theory of Social and Economic Organization.* Translated by A. M. Henderson and Talcott Parsons. New York: The Free Press, 1964.

Weibe, Robert H. *The Segmented Society: An Introduction to the Meaning of America.* New York: Oxford University Press, 1975.

Weigley, Russell F. *The American Way of War: A History of United States Military Strategy and Policy.* New York: Macmillan, 1973.

———. *Towards an American Army: Military Thought from Washington to Marshall.* New York: Columbia University Press, 1962.

Weinstein, James. *The Corporate Ideal in the Liberal State, 1900–1918.* Boston: Beacon Press, 1968.

Werking, Richard Hume. *The Master Architects: Building the United States Foreign Service, 1890–1913*. Lexington: University Press of Kentucky, 1977.

Willoughby, William Franklin. *Government Organization in War Time and After: A Survey of the Federal Civil Agencies Created for the Prosecution of the War*. New York: D. Appleton, 1919.

———. *Principles of Legislative Organization and Administration*. Washington, D.C.: The Brookings Institution, 1934.

Wolk, Herman S. *Planning and Organizing the Postwar Air Force, 1943–1947*. Washington, D.C.: Office of Air Force History, 1984.

Yergin, Daniel. *Shattered Peace: The Origins of the Cold War and the National Security State*. Boston: Houghton Mifflin, 1977.

Yoshpe, Harry B. "Bernard M. Baruch: Civilian Godfather of the Military M-Day Plan." *Military Affairs* 29 (1965): 1–15.

Zacharias, Ellis M. *Behind Closed Doors: The Secret History of the Cold War*. New York: G. P. Putnam's, 1950.

Index

Eberstadt and Forrestal was composed into type on a Compugraphic digital phototypesetter in ten point Galliard with two points of spacing between the lines. Galliard was also selected for display. The book was designed by Jim Billingsley, typeset by Metricomp, Inc., printed offset by Thomson-Shore, Inc., and bound by John H. Dekker & Sons, Inc. The paper on which this book is printed carries acid-free characteristics for an effective life of at least three hundred years.

TEXAS A&M UNIVERSITY PRESS : COLLEGE STATION